Applications of NMR Spectroscopy to Problems in Stereochemistry and Conformational Analysis

Methods in Stereochemical Analysis

Volume 6

Series Editor: **Alan P. Marchand**
Department of Chemistry
North Texas State University
Denton, Texas 76203

Distribution: VCH Verlagsgesellschaft mbH, P.O. Box 1260/1280. D-6940 Weinheim, Federal Republic of Germany

USA and Canada: VCH Publishers, Inc., 303 N.W. 12th Avenue, Deerfield Beach, FL 33442–1705, USA

Applications of NMR Spectroscopy to Problems in Stereochemistry and Conformational Analysis

Edited by
Yoshito Takeuchi and
Alan P. Marchand

Yoshito Takeuchi,
Department of Chemistry
The University of Tokyo
Tokyo 153, Japan

Alan P. Marchand
Department of Chemistry
North Texas State University
Denton, Texas 76203

Library of Congress Cataloging-in-Publication Data
Main entry under title:

Applications of NMR spectroscopy to problems in
stereochemistry and conformational analysis.

(Methods in stereochemical analysis; v. 6)
Includes bibliographies and index.
1. Nuclear magnetic resonance spectroscopy.
2. Stereochemistry. 3. Conformational analysis.
I. Marchand, Alan P. II. Takeuchi, Yoshito, 1934–
III. Series.
QD96.N8A645 1985 541.2′23 85-26291
ISBN 0-89573-118-5

Printed in the United States of America.

ISBN 0-89573-118-5 VCH Publishers
ISBN 3-527-26145-1 VCH Verlagsgesellschaft

**To the memory of
Alvin R. Marchand**

Preface

Few instrumental techniques and developments have had the dramatic impact upon progress in stereochemical and structural analysis that has been evidenced by nuclear magnetic resonance (NMR) spectroscopy. Recent advances in hardware (high sensitivity, high magnetic field instrumentation) and in software (data acquisition, development of new pulse sequences in Fourier transform nuclear magnetic resonance (FT-NMR) spectroscopy, and the like) have revolutionized the field and its applications to structural problems of interest to organic and bio-organic chemists. NMR elucidation of conformation and stereochemistry in high molecular weight polymers and biopolymers, both in solution and in the solid state, is now routinely pursued; often, nuclei that possess low sensitivity and/or low natural abundance are employed for such studies. The advent of two dimensional (2-D) NMR spectroscopy has had a profound effect upon NMR applications; molecules whose NMR spectra had been regarded only a few years ago as being intractably complex are now routinely analyzed by the use of 2-D NMR techniques.

As a result of these and related developments, there has been an explosion of interest in applications of NMR spectroscopy to structural and stereochemical problems. It is certainly appropriate to state that maintaining current awareness of developments in this burgeoning interdisciplinary field is vitally important to chemists and biochemists whose research programs are concerned with the determination of gross molecular structure as well as with details of molecular stereochemistry and conformational analysis. It is in this spirit that the preparation of the present volume was undertaken.

Given the scope of recent developments in NMR spectroscopy, it did not seem reasonable to attempt an exhaustive exposition of the subject in a single volume. In addition the interdisciplinary nature of NMR applications virtually mandated that we approach this subject through a multiauthored treatise format. To this end, we sought contributions from diverse segments of the chemical community.

In the first chapter Professor I. O. Sutherland (University of Liverpool) presents new NMR applications to the study of conformational analysis of guest-host molecular systems, many of which are of intense current interest to investigators in the life sciences. In Chapter 2 Professor P. Diehl (University of Basel) leads us through the intricacies of NMR investigations in liquid crystal media and demonstrates how the study of partially oriented molecules can provide information pertaining to their molecular geometry. In Chapter 3 Professor R. Kitamaru (Kyoto University) illustrates the use of ^{13}C relaxation phenomena as a source of information pertaining to chain dynamics and conformation in macromolecules. In the fourth chapter Professor T. Terao

(Kyoto University) examines the use of NMR spectroscopy to study stereochemical problems in the solid state. Two-dimensional NMR spectroscopy is treated in the last two chapters: in Chapter 5 Professor K. Nagayama (University of Tokyo) introduces the basic concepts of 2-D NMR spectroscopy and presents a number of applications of 2-D NMR techniques; in the final chapter Professor H. Kessler (Johann Wolfgang Goethe University, Frankfurt) examines problems in conformational analysis of peptides using 2-D NMR techniques.

The editors acknowledge with pleasure the contributions and suggestions made by numerous colleagues and by the individual authors whose names appear above. We are especially grateful for the support and encouragement which we received from members of our respective families. We thank Mrs. Rosalind B. Marchand for assistance in preparing the Author Index. Finally, we thank Ms. Mary Stradner, Marie Stilkind, and other persons associated with the Production Division of VCH Publishers, Inc.

Yoshito Takeuchi
Tokyo, Japan

Alan P. Marchand
Denton, Texas

CONTENTS

1

GUEST–HOST CHEMISTRY AND CONFORMATIONAL ANALYSIS

Ian O. Sutherland

Department of Organic Chemistry
The Robert Robinson Laboratories, The University of Liverpool
P.O. Box 147, Liverpool, L69 3BX

Introduction

Conformational analysis in organic chemistry has generally been based upon the conformational preference of molecules in their ground states and the stereochemical requirements of transition states. Ground state conformations are usually considered for isolated molecules and the conformational analysis of transition states is based upon the formation of more or less well-defined partial bonds between reacting centers in either a single molecule or a pair of molecules. Such considerations[1] have played an important part in the appreciation of structure and reactions in organic chemistry over the last three decades, since the publication of the seminal paper by Barton[2] in 1950. The development of guest–host chemistry[3-7] over the past 10 years has focused attention on systems involving two or more molecules, bound together in a noncovalent fashion. This requires the introduction of methods for studying the conformations of multimolecular systems or intermolecular conformational analysis. Information about the solid phase is, of course, readily available through x-ray crystallography, but information for multimolecular systems in solution must be obtained by more indirect methods; this chapter reviews the application of NMR spectroscopy to the study of intermolecular conformational analysis. This chapter is restricted to cases in which the lifetime of the complex is

NMR in Stereochemical Analysis

significant (τ ca. 1 s) on the NMR time scale at a temperature within the range ($-110°C$ to $+100°C$) over which complexes may be studied in solution by NMR spectroscopy. These limits are somewhat arbitrary and they exclude, for example, collision complexes and charge-transfer complexes with very short lifetimes, as well as strongly bound metal–ligand complexes with very long lifetimes.

The range of lifetimes selected above is of the same order as that found in biological processes involving complex formation, and one of the incentives for the study of guest–host chemistry is that of providing analogs of biological systems. Examples are the modification of a substrate, $S \rightarrow S'$, by an enzyme E (eq. 1-1, where the formation of a complex is indicated by $E \cdot S$ and $E + S$ indicates a mixture of the free components); the interaction of a substrate S, such as a drug, with a biological receptor R to initiate a biological response (eq. 1-2); or even the transport of a substrate S through a cell membrane or across a phase boundary by the formation of a complex with a carrier molecule C (eq. 1-3, where the vertical lines represent phase boundaries).

$$E + S \leftrightharpoons E \cdot S \leftrightharpoons E \cdot S' \leftrightharpoons E + S' \qquad (1\text{-}1)$$

$$R + S \rightarrow R \cdot S \rightarrow \text{biological response} \qquad (1\text{-}2)$$

$$S + |C \leftrightharpoons C \cdot S \leftrightharpoons C| + S \qquad (1\text{-}3)$$

These processes summarized here are related to other important biological systems, including immune response and the biological activity of antibiotics, toxins, hormones, and neurotransmitters. The discussion in this chapter, however, is centered upon the systems involving synthetic host molecules.

The binding energy required for the formation of a complex of the type to be discussed is derived primarily from (a) coulombic attraction, (b) hydrogen bonding, and (c) hydrophobic interactions. The first two of these binding forces are possible in all types of solvent but are particularly favored in organic solvents of low dielectric constant; the third binding force is restricted to aqueous systems. The best studied complexes in organic solvents are those formed between cations and crown ethers, whereas systems that have been studied in aqueous solvents include complexes of cyclodextrins[8-10] and of water-soluble cyclophanes.[11-13] Before specific examples of these systems are discussed, it is worth noting that a number of natural products besides proteins owe their biological activity to their ability to function as host molecules. The first group of these, the ionophoric antibiotics,[14] function as hosts for the simplest guests of all, spherical metal cations. Information on the conformational aspects of complexation generally has been derived from crystal structures of free and complexed host molecules, rather than from NMR studies. The more complex group of peptide-like antibiotics, exemplified by vancomycin,[15-18] ristocetin,[19,20] and avoparcin,[21,22] form complexes selectively with *N*-acyl-D-Ala-D-Ala derivatives, and complexation has been studied by NMR techniques.

The study of the binding site of vancomycin is a particularly interesting illustration of the power of NMR methods. The binding site originally deduced from the crystal structure[23] of a vancomycin degradation product was subsequently revised on the basis of NMR spectroscopy and further chemical studies.[16,17,18] In the original proposal[23] the NH and CO groups enclosed in rectangles in **1** were used as the basis for hydrogen bonding to an *N*-acetyl-D-Ala-D-Ala guest. A subsequent examination of the ^1H NMR spectrum of the vancomycin–acetyl-D-Ala-D-Ala complex under conditions of slow exchange between the complex and its free components [$-1°C$ in $(CD_3)_2SO–CCl_4$] showed that the proton of the circled NH group in **1** moved downfield 3.7 ppm in the complex as compared with free vancomycin. Therefore, this NH group is involved in hydrogen bonding to the free carboxylate group of the substrate. From a study of molecular models it was recognized that this implied a major conformational change of the right-hand side of structure **1** in the vancomycin complex to provide a carboxylate anion "receptor pocket" in the vancomycin structure. This receptor pocket was found from the models to resemble closely the receptor pocket found for the related antibiotic ristocetin **2**. Therefore, the

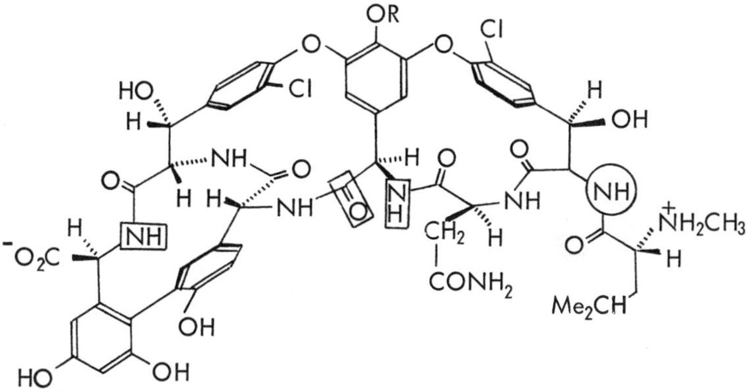

R = sugars

1

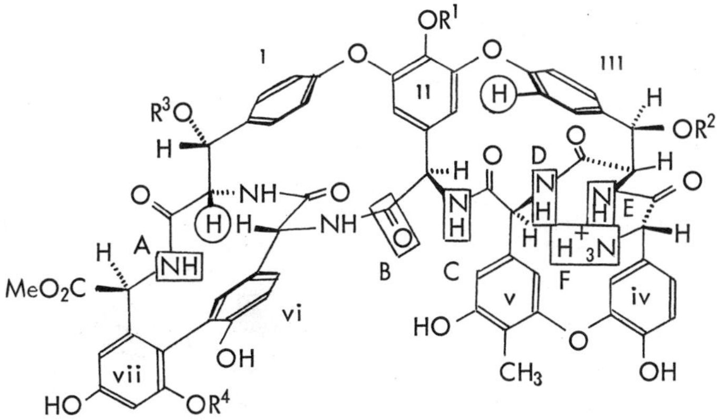

R^1, R^2, R^3, R^4 = sugars

2

NH, CO, and $\overset{+}{N}H_3$ groups in ristocetin enclosed in rectangles (see **2**) have been shown,[20] on the basis of molecular models and *intermolecular* nuclear Overhauser enhancements (NOEs) between the encircled protons of the host antibiotic (see **2**) and the C*H* and CH_3 protons of the D-Ala residues of the guest, to be placed correctly for the formation of intermolecular hydrogen bonds. The diagrammatic representation (**3**, where the binding sites are labeled A–F and the aromatic rings i–vii to correspond with the labels in **2**) of this hydrogen bonding between host and guest shows all of the postulated binding interactions. A similar set of hydrogen bonds can be found for an appropriate conformation of vancomycin. The *intramolecular* NOEs found for both vancomycin and ristocetin and their complexes support these views regarding the conformational similarities of the macrotricyclic host, vancomycin **1**, and the macrotetracyclic host, ristocetin **2**. The related antibiotic avoparcin[21,22] has a macrotricyclic structure **4** that presumably adopts a conformation similar to the conformations of vancomycin and ristocetin.

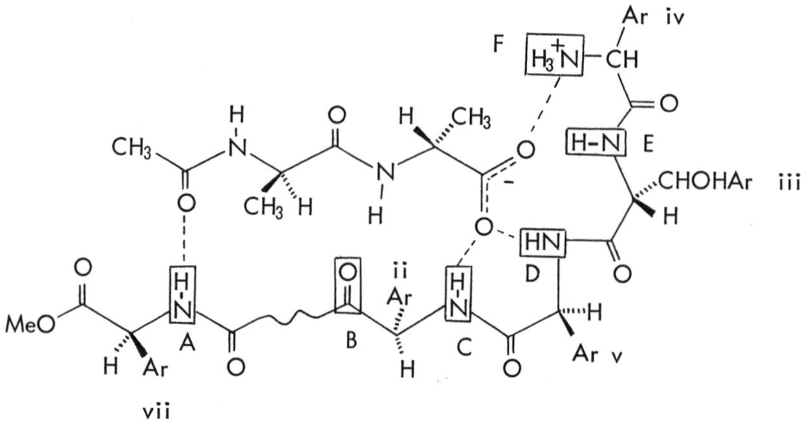

3

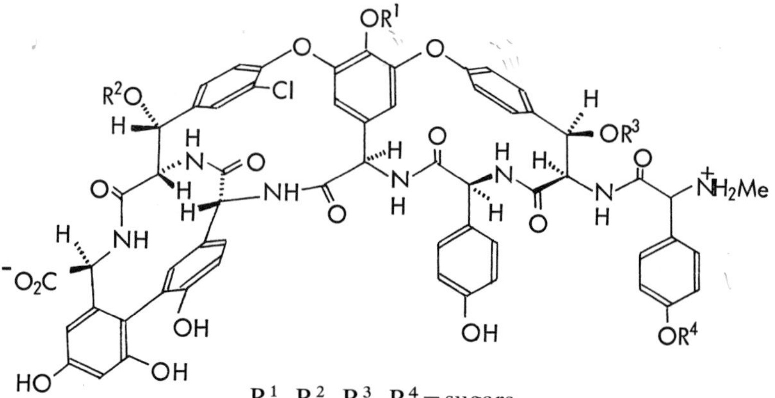

R¹, R², R³, R⁴ = sugars

4

These studies of conformation and intermolecular relationships using observations of intramolecular and intermolecular NOEs are of great interest and are likely to serve as models for other investigations of guest–host complexes involving organic guest molecules. They are most applicable, however, to asymmetrical systems having well-dispersed ^1H NMR spectra, whereas many synthetic host molecules are symmetrical; and in general, the NOE has not yet been applied for the study of complexes of such symmetrical host molecules.

Complexes of Crown Ethers and Analogous Hosts

The discovery by Pedersen[24,25] that crown ethers, such as 18-crown-6 **5**, form strongly bound complexes with alkali metal and other metal cations and alkylammonium cations has generated an impressive volume of research. The complexation of metal cations by crown ethers is an important area of research, but in many cases the structures of such complexes have been studied by x-ray crystallography, rather than by NMR spectroscopy. The study of metal ion complexation by cryptands, crown ethers, and ionophoric antibiotics by a variety of techniques has been reviewed;[26] it is not discussed further in this chapter. The complexation of alkylammonium cations has proved to be particularly suitable for study by NMR techniques, and an account of this work constitutes the bulk of this section.

Monocyclic Systems

Methods. Early studies of crown ether complexes by ^1H NMR spectroscopy were limited to the recognition of the changed chemical shifts for both guest and host molecules upon complexation[27] and to the measurement of the relative amounts of guest and host species present in solution in order to determine values for association constants.[27-29] It was noted in this early work that guest–host exchange was fast on the NMR time scale at normal probe temperatures, resulting in, for example, the averaging of signals for diastereoisomeric complexes simultaneously present in solution. At that time (1973), although slow exchange had been recognized for complexes of macromolecular

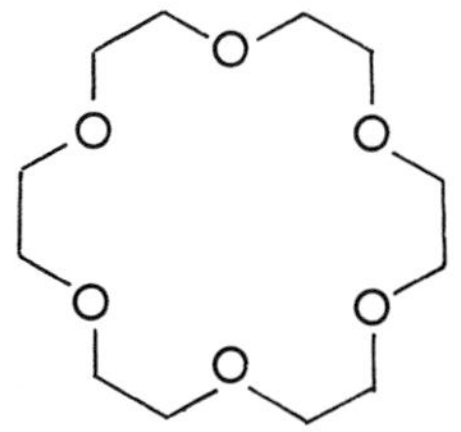

O O O O O O

5

systems, there was no well-defined example of a hydrogen-bonded complex between molecules of relatively low molecular weight in which exchange was slow on the NMR time scale, even at very low temperatures. It was, therefore, surprising to find that complexes of the metacyclophanes **6** with benzyl-ammonium thiocyanate in CD_2Cl_2 showed temperature dependence of their ^1H NMR spectra that was not matched by that found for the free host molecules.[30] The indicated protons of the benzylic CH_2 groups of the hosts **6** complexed with benzyl-ammonium thiocyanate gave singlet signals, at ambient temperatures, which changed to AB systems at low temperatures (for **6**, $n=2$, at $<-45°C$ and for **6**, $n=3$, at $<-65°C$). This change was consistent with the formation of a complex (in **7** and a number of other formula of this type the crown ether system is represented by a circle or part of a circle for simplicity) in which exchange of the guest cation between the two faces of the host **7a**⇋**7b** became slow on the NMR time scale at low temperatures. The free energy of activation associated with this process could be obtained in the usual way[31-33] from the coalescence data for the spectroscopic change. The process **7a**⇋**7b** is clearly associated with both face to face exchange of the guest molecule (E) and an appropriate conformational change (I) of the host molecule. This conformational change must involve the interconversion of conformers related by a horizontal plane of symmetry; it is conveniently described as "conformational inversion" (I) and has an obvious similarity to the conformational inversion of, for example, six-membered ring systems. The overall process $E+I$, **7a**⇋**7b**, may or may not involve complete dissociation of the complex.

These early results were extended in later work, and a number of different

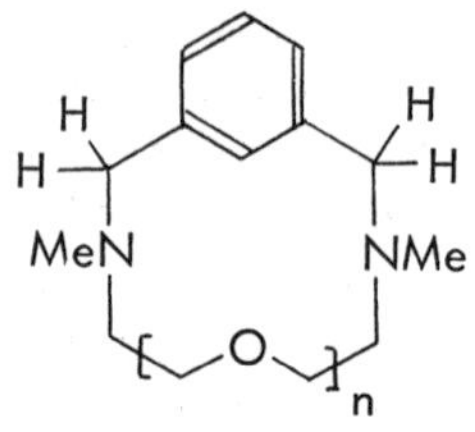

$n=2$ or 3
6

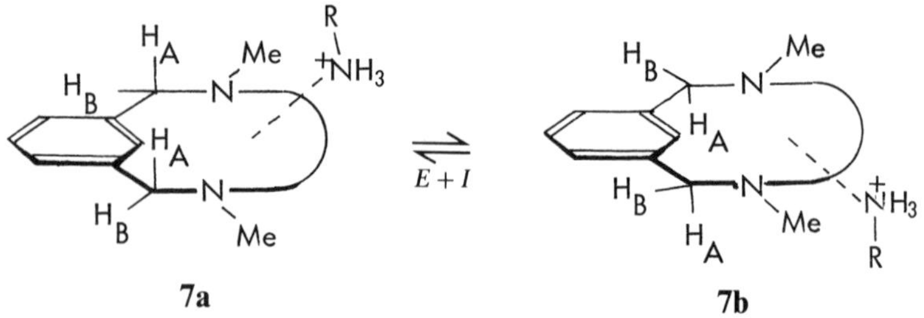

$E+I$

7a **7b**

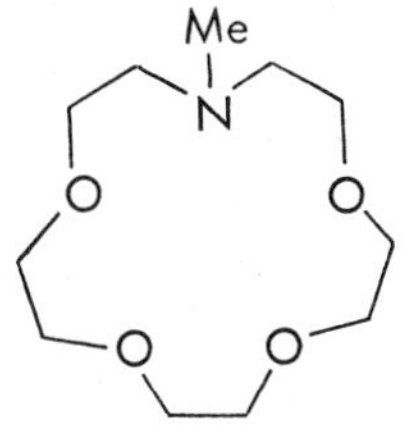

8

types of dynamic behavior were recognized for crown ether–alkylammonium cation complexes for which structural evidence relevant to the complexation could be derived.[34] This is best exemplified by a detailed study[35] of the complexes of the monoaza-15-crown-5 derivative **8**. The ^1H NMR spectrum of the host **8** is unexceptional at 25°C and does not show any evidence for slow conformational changes in the temperature range +25° to −110°C. The NMR spectrum of the 1:1 complex of **8** and benzyl-ammonium thiocyanate in CD_2Cl_2 shows changed chemical shifts for the host protons, and the NCH_2 signal, observable as a triplet at 25°C, changes below −22°C to two broad multiplets that are assignable to H_A and H_B of the four-spin system $NCH_AH_BCH_2O$. This change is consistent with face to face exchange of the guest cation (**9a**⇆**9b**, where the labels A, B, C, D, etc, are used to denote hydrogen atoms in environments A, B, C, D, etc) becoming slow on the NMR

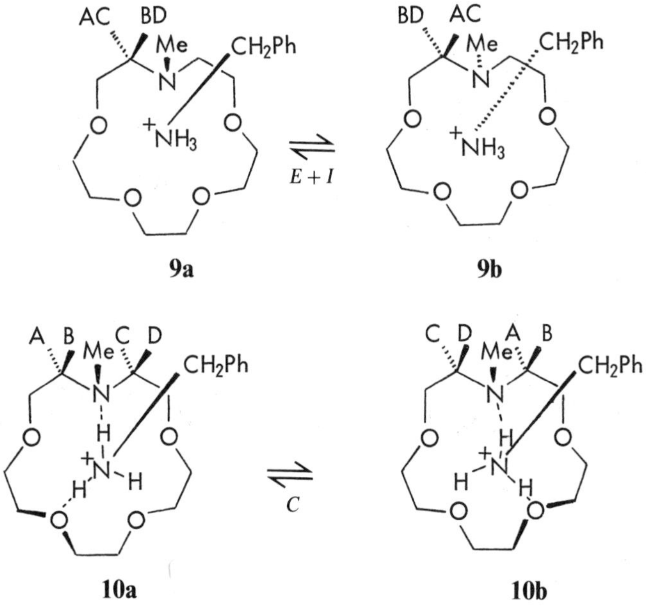

9a ⇌ (*E* + *I*) **9b**

10a ⇌ (*C*) **10b**

Scheme I

time scale. Finally, at very low temperatures ($< -102°C$) the NCH_2 multiplets separate to give two pairs of signals assignable to H_A, H_B, H_C, and H_D of the slowly interconverting species **10a** and **10b**. The process **10a**$\rightleftharpoons$**10b** (C) involves reorganization of the hydrogen bonding and, probably more importantly, rotation about an OCH_2—CH_2O bond of the macrocycle from one gauche conformation to the other. The site exchanges associated with these interconversions $E+I$ **9a**$\rightleftharpoons$**9b** and C **10a**$\rightleftharpoons$**10b** are summarized in Scheme I.

These assignments of NMR temperature dependence to rate processes were checked by studying the complex of the host **8** with a chiral guest, 1-phenylethylammonium thiocyanate. The complex with the (R)–guest (1:1 ratio) gave a multiplet for the NCH_2 protons at 30°C, corresponding to a fast rate for the processes $E+I$ and C. At lower temperatures ($< -20°C$) the process $E+I$

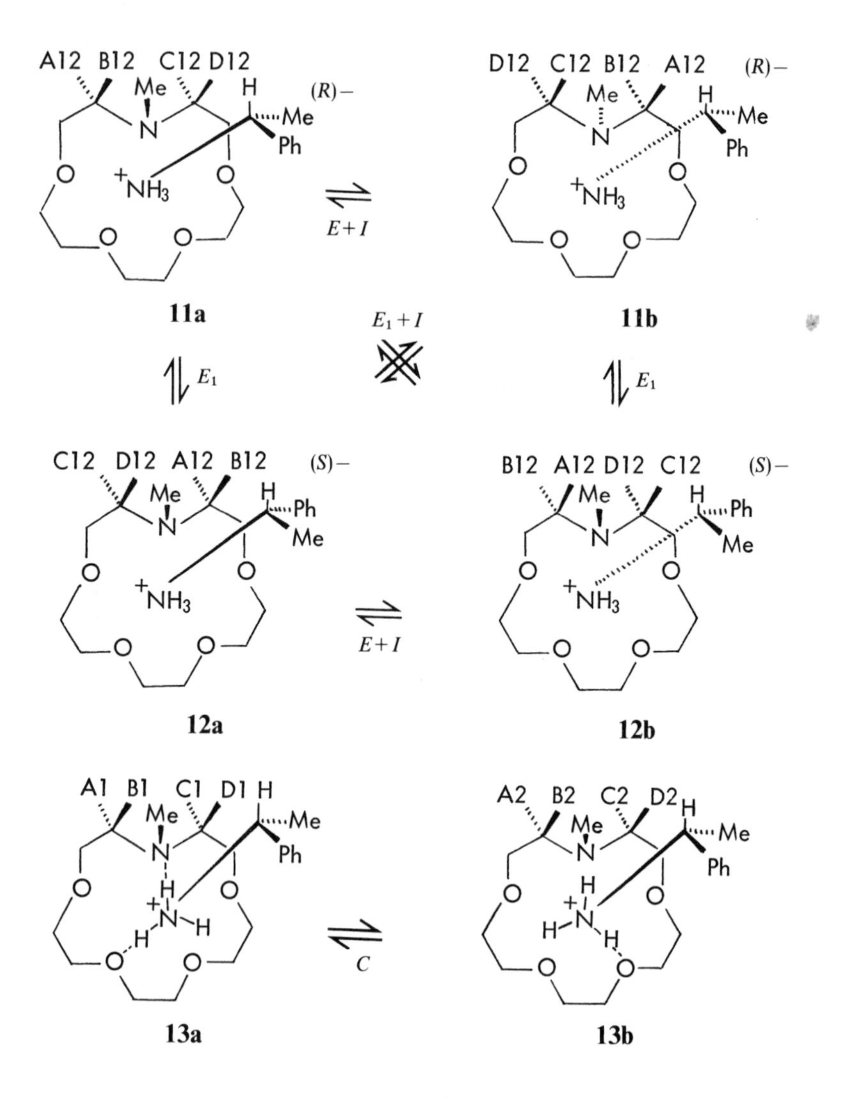

11a$\leftrightarrows$**11b** became slow on the NMR time scale, leading to the observation of four separate multiplet signals for the protons labeled A12, B12, C12, and D12 in **11a** and **11b** (where the descriptions A12, B12 etc, imply site averaging for the pairs of sites A1 and A2, B1, and B2, etc; similarly, the description AB12 implies averaging of the four sites A1, A2, B1, and B2 and ABCD12 for the eight sites A1, A2, B1, B2, C1, C2, D1, and D2). Finally, at very low temperatures ($< -95°C$) further complex changes were observed in the NMR spectrum as a result of slow interconversion of the diastereoisomeric species **13a** and **13b** (the assignment of sites to the diastereoisomers **13a** and **13b** is arbitrary), leading to the direct observation of four broad signals assignable to the NCH_2 protons (A1, B1, C1, and D1) of the major species and the implication of four further signals for the minor species (A2, B2, C2, and D2). This final change also affects the *N*-Me signals, the major species giving an observable high-field singlet and the minor species an implied singlet at lower field. The temperature dependence of the spectrum of the complex of **8** with (*R,S*)-phenylethylammonium thiocyanate is different; this is expected considering the additional possibilities for site exchange associated with the processes **11a**$\leftrightarrows$**12a** and **11b**$\leftrightarrows$**12b** (E_1), and **11a**$\leftrightarrows$**12b** and **11b**$\leftrightarrows$**12a** (E_1+I) involving exchange of guest molecules of differing chirality. Hence, at 30°C all site exchanges are fast and a single triplet is observed for the host NCH_2 protons; at lower temperatures ($< -30°C$) E_1 is fast but ($E+I$) and (E_1+I) are slow, leading to the observation of a pair of broad multiplets (AC12 and BD12); finally at $-60°C$ E_1 also becomes slow and four signals are observable, corresponding to the averaged sites A12, B12, C12, and D12. Below this temperature the changes are identical to those observed for the complex of the (*R*)-guest. The site exchanges included in the above discussion are summarized in Figure 1-1. It should be noted that the whole discussion is based upon complexes in which the guest cation is situated in a cis relationship with respect to the *N*-Me substituent of the host; this is consistent with the high-field shift of the NCH_3 protons due to the shielding effect of the phenyl substituent in the guest cation.

If solutions of the complexes are prepared using an excess of the guest salt, the ^1H NMR spectra of the solutions at low temperatures (E slow) show separate signals corresponding to free and complexed guest cations. At higher temperatures signal averaging occurs as expected for fast exchange between the free and complexed species. Activation energies ($\Delta G^\ddagger$) may be obtained for this process and for all the other rate processes based upon signal separation and coalescence temperatures.[31] Values obtained for these activation energies are summarized in Table 1-1.

The related monoaza-18-crown-6 derivative **14** shows similar spectroscopic characteristics, but in addition the ^1H NMR spectrum of a 1:1 host–guest ratio shows signals for two diastereomeric complexes at low temperatures. It is reasonable to assume that these are related as the cis and trans species **15** and **16**. The interconversion **15**$\leftrightarrows$**16** (E) involves a face to face exchange of the guest cation without total inversion of the host macrocycle. The energy barrier

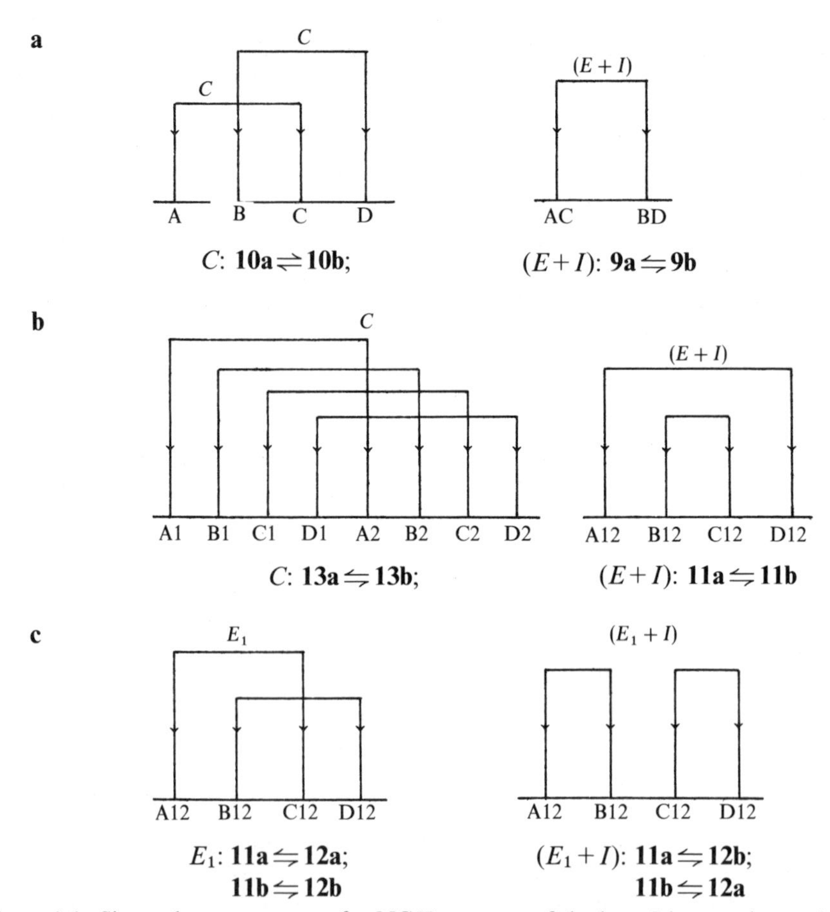

Figure 1-1. Site exchange processes for NCH_2 protons of the host **8** in complexes with (**a**) benzylammonium thiocyanate, (**b**) (*R*)-phenylethylammonium thiocyanate, and (**c**) (*R,S*)-phenylethylammonium thiocyanate.

associated with this change (Table 1-1) is of the same order as that determined for the exchange of free and complexed guest cations, and it is believed that both processes involve dissociation of the complex as the rate-determining step.

The use of a chiral guest as a single enantiomer, (*R*)-PhCHMe$\overset{+}{N}H_3$ NCS^-, allows the processes E and $E+I$ to be examined in the same way as for the analogous complex of the monoaza-15-crown-5 system **8**, with the additional complexity resulting from the presence of both cis and trans complexes. The use of a racemic chiral guest, (*R,S*)-PhCHMe$\overset{+}{N}H_3$ NCS^-, leads to the two further possibilities for guest exchange exemplified by **17**⇋**18** (E_2) and **18**⇋**19** (E_1); both of these additional processes are found to have free energies of activation ($\Delta G^{\ddagger}$) of the same order as that associated with the exchange E. The energy barriers for exchange processes derived from examination of the complexes of **14** are summarized in Table 1-1.

TABLE 1-1. FREE ENERGIES OF ACTIVATION ($\Delta G^{\ddagger}$) FOR PROCESSES INVOLVING GUEST-HOST EXCHANGE AND CONFORMATIONAL CHANGES FOR COMPLEXES OF HOSTS **8** AND **14** WITH PRIMARY ALKYLAMMONIUM THIOCYANATES, $R{-}\overset{+}{N}H_3 \cdot NCS^-$ IN CD_2Cl_2

Host	Guest, R	Ratio, H:G	Signal	Process	$\Delta G^{\ddagger}$ (kcal mol^{-1})
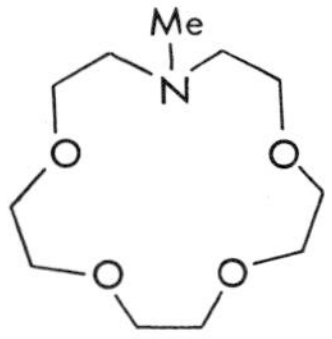 **8**	$PhCH_2$	1:1	NCH_2	$E+I$	12.3
			NCH_2	C	8.0
	(R)-PhCHMe	1:1	NCH_2	$E+I$	12.3
			NCH_3	C	≈8.4
	(R,S)-PhCHMe	1:2	NCH_2	$E+I$, E_1+I	12.1
			$PhCHCH_3\overset{+}{N}H_3$	E, E_1	10.2
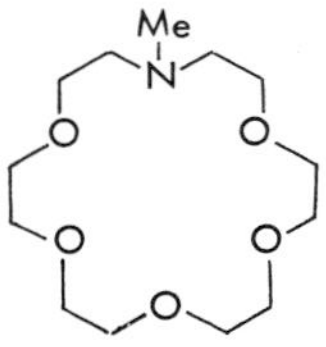 **14**	$PhCH_2$	1:1	NCH_2	$E+I$	12.6
			NCH_3	E	10.7
	$PhCH_2$	1:2	NCH_2	$E+I$	12.7
			NCH_3	E	9.9
	(R)-PhCHMe	1:1	NCH_2	$E+I$	12.2
			NCH_3, $\overset{+}{N}H_3$	E	10.7
	(R)-PhCHMe	1:2	NCH_3	E	10.5
			$PhCHCH_3\overset{+}{N}H_3$	E	10.2
	(R,S)-PhCHMe	1:2	NCH_2	$E+I$, E_1+I, E_2+I	12.1
			$PhCHCH_3\overset{+}{N}H_3$	E, E_1, E_2	10.2

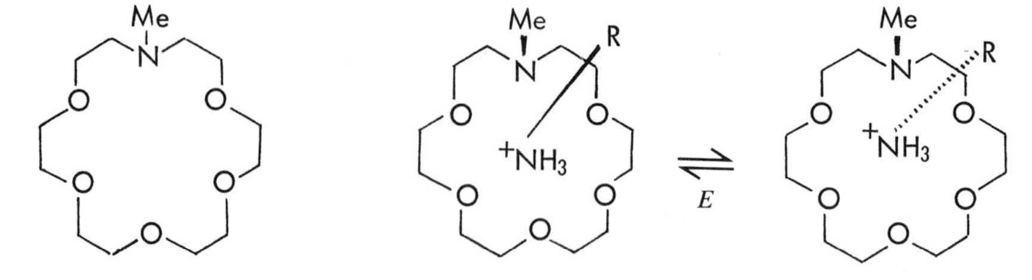

14 15 16

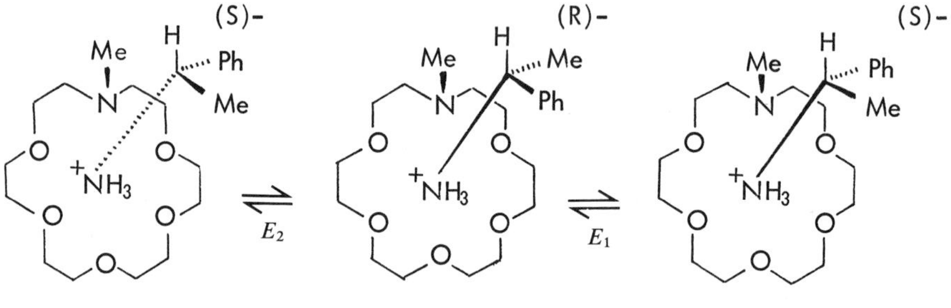

17 18 19

Processes similar to those described for the above complexes of the monoaza crown ethers **8** and **14** also have been found for complexes of 18-crown-6 **5** and for a variety of diaza crown ethers.[30,34,36-38] In general, three types of process are detectable.

1. **Guest exchange.** This includes exchange of free and complexed guest cations without a major conformational change of the host molecule. The energy barriers associated with these processes may depend, in some cases, upon the concentration of the guest cation, suggesting a bimolecular mechanism for the exchange process. In most cases involving aza crown systems, exchange probably involves dissociation of the complex to give the free components followed by association; the free energy profile for such a process is summarized in Figure 1-2a.
2. **Guest exchange with inversion of the host conformation.** Processes of this type involve a face to face exchange of the guest cation and the interconversion of complexes of similar type. They are recognizable by the collapse of AB systems associated with NCH_2 and OCH_2 groups of the macrocycle as the process becomes fast on the NMR time scale. Processes of this type probably proceed by the sequence, dissociation, conformational change of the free host molecule, and association. The free energy profile for such a process is given in Figure 1-2b, which shows how the process may be associated with a higher activation energy than the simple guest exchange process.
3. **Conformational changes within the complex.** These processes are recogniz-

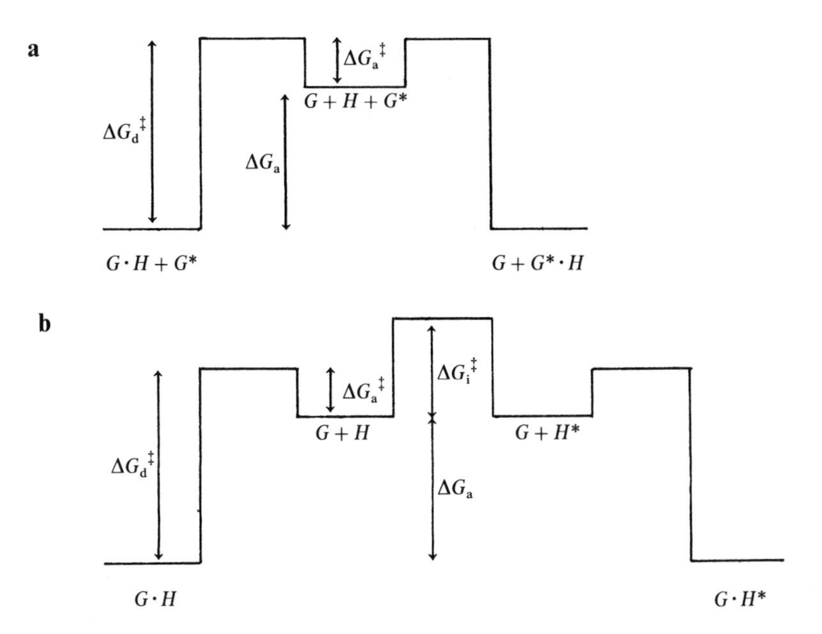

Figure 1-2. Free energy profiles for (**a**) guest exchange and (**b**) guest exchange with inversion of the host conformation. ΔG_a is the free energy of complexation, $\Delta G_a^\ddagger$ and $\Delta G_d^\ddagger$ are the activation free energies for association and dissociation, and $\Delta G_i^\ddagger$ is the activation free energy for conformational inversion of the host macrocycle.

able when they involve activation energies that are significantly lower than those associated with guest exchange. They appear to be general for strongly bound complexes where complexation can involve two enantiomeric or diastereoisomeric conformations of the host macrocycle and where pairs of complexes involving these conformations can be interconverted by a process that does not require dissociation of the complex. Such processes may involve rotation from one gauche conformation to another about OC—CO and NC—CO bonds and possibly reorganization of hydrogen bonding at a single face of the host molecule. The energy barrier for such a process has no direct relationship with the activation energy for dissociation, $\Delta G_d^\ddagger$, or the free energy, ΔG_a, of complexation.

The energy profiles in Figure 1-2 show that $\Delta G_d^\ddagger$, the activation energy for dissociation of the complex, and $\Delta G_i^\ddagger + \Delta G_a$, the sum of the activation energy for inversion and the free energy of complexation, may be equated directly to the free energies of activation for processes of types 1 and 2, respectively. Therefore, the study of the dynamic behavior of complexes may provide a guide to the relative free energies of complexation. The dynamics of guest–host exchange have also been related to the free energy of complexation in elegant NMR studies described by de Jong and Reinhoudt and their collaborators.[39,40] These investigators examined the exchange of a guest *t*-butylammonium cation between two crown ethers, I and II. It was shown that exchange was normally fast at room temperature, but at lower temperatures the CMe_3 signal broadened, and eventually peak separation was observed. In general, three environments for the *t*-butylammonium cation AM are expected: (a) the complex with crown ether I, I·AM; (b) the complex with crown ether II, II·AM; and (c) the free cation, AM. Consideration of the equilibria for dissociation of the two complexes gives the following relationships:

$$\mathrm{I{\cdot}AM} \underset{k_{-1}}{\overset{k_1}{\rightleftharpoons}} \mathrm{I} + \mathrm{AM}; \ K_1 = k_{-1}/k_1 \tag{1-4}$$

$$\mathrm{II{\cdot}AM} \underset{k_{-2}}{\overset{k_2}{\rightleftharpoons}} \mathrm{II} + \mathrm{AM}; \ K_2 = k_{-2}/k_2 \tag{1-5}$$

where K_1 and K_2, are the equilibrium constants and k_1, k_{-1}, k_2, and k_{-2} are the rate constants for the formation and dissociation of the two complexes. For a case in which $k_1 \ll k_2$, k_1 is rate determining for cation exchange between the two complexes and can be found by applying the simple two-site line shape equation to the CMe_3 signals from the two complexes provided that a temperature can be attained at which k_1 is slow on the NMR time scale. The possibility of the bimolecular contribution to the exchange, summarized in eq. (1-6), was also considered:

$$\mathrm{I{\cdot}AM} + \mathrm{AM}^* \xrightarrow{k_3} \mathrm{I{\cdot}AM}^* + \mathrm{AM} \tag{1-6}$$

If this contribution is significant, then the measured rate constant k is a function of k_1 and k_3:

$$k = k_1 + k_3\,[\mathrm{AM}] \tag{1-7}$$

Exchange rates for the case where I is 18-crown-6 **5**, II is the 18-crown-5 derivative **20**, and AM is *t*-butylammonium hexafluorophosphonate were measured for two different conditions: (1) measurements were made for $CHCl_3$ solutions containing I·AM and II·AM only; under these conditions, for

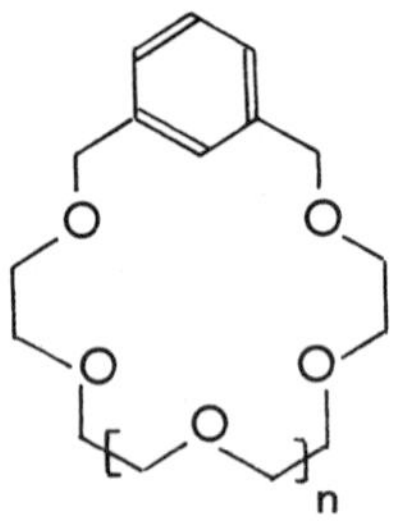

20, $n=1$
21, $n=2$

$k_1 \gg k_2$, the concentration of the free salt AM is given by:

$$[AM] = ([II \cdot AM]/K_2)^{\frac{1}{2}} \tag{1-8}$$

A plot of k against $[II \cdot AM]^{\frac{1}{2}}$ gives k_1 by extrapolation to $[II \cdot AM] = 0$ and the slope is $k_3/K_2^{\frac{1}{2}}$.

(2) Measurements were also made for $CHCl_3$ solutions containing the complexes and an excess of host II (concentration $[II]_e$); under these conditions the concentration of the free salt AM is given by:

$$[AM] = [II \cdot AM]/K_2[II]_e \tag{1-9}$$

A plot of k against $[II \cdot AM]/[II]_e$ gives k_1 by extrapolation to $[II \cdot AM]/[II]_e = 0$, and the slope is k_3/K_2.

Application of this method gave the following values of rate constants and equilibrium constants and associated activation parameters, E_a, for the complexes of 18-crown-6 **5** and the 18-crown-5 derivative **20** at 20°C in $CHCl_3$:

$$k_1 = 70\ s^{-1},\ E_a = 20.4 \pm 1\ \text{kcal mol}^{-1},\ K_2 = 2.2 \times 10^5\ M^{-1}$$

hence

$$K_1 = 2.3 \times 10^7\ M^{-1}$$

$$k_{-1} = 1.6 \times 10^9\ s^{-1}$$

$$k_3 = 1{\cdot}3 \times 10^6\ M^{-1}\ s^{-1},\ E_a = 8.1 \pm 1\ \text{kcal mol}^{-1}$$

Therefore, bimolecular exchange of the type described in eq. (1-6) has a low activation energy and makes a contribution to k; it is assumed to involve an intermediate 2:1 complex, AM*·I·AM.

This rigorous approach (method A) to the determination of the kinetics of complexation was compared[40] with rather less rigorous approaches based upon the rate of face to face exchange of the complexed cation (method B) as determined from the coalescence of the AB system for OCH_2 groups of the macrocyclic host, and the exchange of free and complexed crown ether (method C) for solutions of complexes containing an excess of the host macrocycle. All three approaches gave results in reasonable agreement, although method A was considered to be the most reliable. Some typical results are given in Table 1-2; in general $\Delta G_d^{\ddagger}$ for dissociation, derived from measurement of k_1 at a single temperature, was found to be ca. 5 kcal mol^{-1} greater than $-\Delta G_a$, the free energy of complexation. This suggests that the association process is controlled by diffusion for all of the cases summarized in Table 1-2 [see also Fig. 1-2a].

Subsequently, method B was used to examine[40] the variation in $\Delta G_d^{\ddagger}$ for complexation of the host **20** with a range of primary alkylammonium salts, $R\overset{+}{N}H_3PF_6^-$. Very little variation was found in $\Delta G_d^{\ddagger}$ for a wide range of alkyl groups R of differing steric requirements in contrast with previous studies of aza crown ethers,[34,36] in which $\Delta G_d^{\ddagger}$ was decreased by increasing steric demand of the guest alkylammonium cation. The variation in $\Delta G_d^{\ddagger}$ for complexes of host **20** with the salts $R\overset{+}{N}H_3X^-$ was also studied,[40] using method B, for a range of anions, X^-. The results of this study, summarized in Table 1-3, show that $\Delta G_d^{\ddagger}$ decreases as the hydrogen bonding ability of the anion in apolar solvents increases, the most kinetically stable complexes being formed

TABLE 1-2. FREE ENERGIES OF ACTIVATION ($\Delta G_d^{\ddagger}$) AND FREE ENERGIES OF ASSOCIATION ($-\Delta G_a$) FOR COMPLEXES OF HOSTS **5** AND **20–25** WITH *t*-BUTYLAMMONIUM THIOCYANATE IN $CDCl_3$ AT 20°C

Host	k_1	$\Delta G_d^{\ddagger}$ (kcal mol^{-1})	K_1 (M^{-1})	$-\Delta G_a$ (kcal mol^{-1})
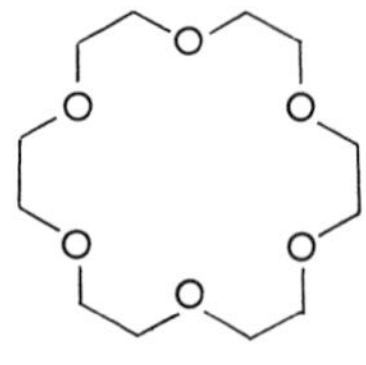 **5**	65	14.7	2.3×10^7	9.8
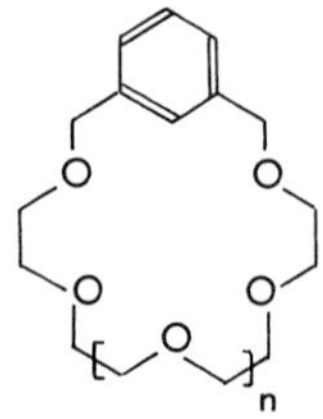 **20**, $n = 1$	5400	12.1	2.2×10^5	7.1
21, $n = 2$			2.9×10^4	5.9
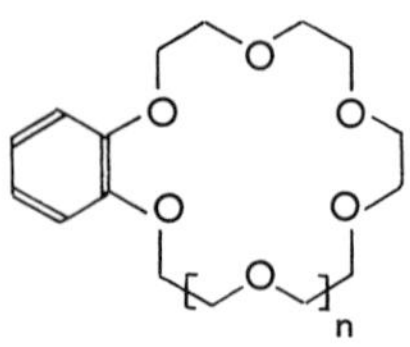 **22**, $n = 0$			8.2×10^3	
23, $n = 1$	155	14.2	5×10^6	8.9
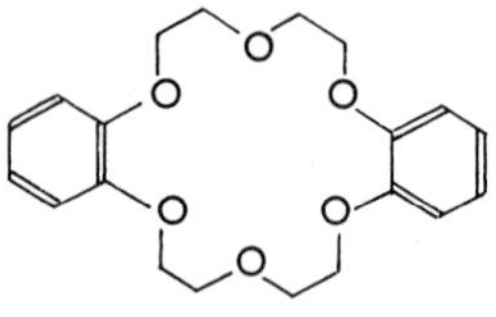 **24**	850	13.2	1.2×10^6	8.1
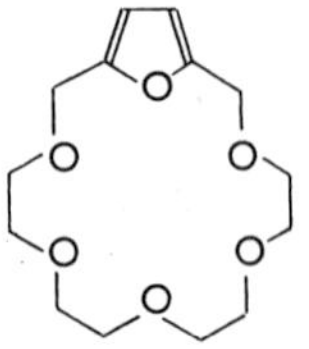 **25**	1100	13.0	1.3×10^6	8.2

TABLE 1-3. VARIATION IN $\Delta G_d^\ddagger$ (kJ mol^{-1}) FOR CHANGES IN X^- IN COMPLEXATION OF AMMONIUM SALTS, $R\overset{+}{N}H_3X^-$, BY HOST **20** (0.025 *M* IN $CDCl_3$)

	R		
X^-	Me_3C	$PhCH_2$	p-MeC_6H_4
PF_6^-	51.5	54.0	54.9
ClO_4^-	46.1	49.8	51.1
BF_4^-		49.0	
I^-		46.8	
SCN^-	<43[a]	45.2	<43[a]
Cl^-	<43[a]	<43[b]	<43[a]

[a] $T_c < -65°C$.
[b] $T_c < -70°C$.

for anions that do not form hydrogen bonds readily. It is probable that ΔG_a responds in a similar way to changes in the hydrogen-bonding ability of the anion. This variation is consistent with hydrogen-bonding between all three hydrogens of the $\overset{+}{N}H_3$ group to the macrocyclic oxygen atoms of the host **20**, with little or no contribution being made to the structure of the complex in $CDCl_3$ solution from hydrogen-bonding to the counter ion. This conclusion also is in accord with the known crystal structure of the complex of a monobromo derivative of the crown ether **20** with *t*-butylammonium thiocyanate.[41,42]

Examples. The NMR methods just discussed have been applied to a wide variety of complexes of monocyclic analogs of crown ethers. A few selected examples, particularly those related to the next part of this chapter, are discussed in this section.

All the studies of oxygen crown systems that have been reported, apart from those just discussed, have been concerned with alkylammonium salt complexes of 18-crown-6 derivatives based upon carbohydrates, with crown dilactones, and with complexes of 18-crown-6 with metal ammines. The carbohydrate-derived crown ethers **26** and **27** are typical of the first type of host molecule;[43] they present two different faces for complexation of an ammonium salt, and hence each of these hosts forms two diastereoisomeric complexes with a guest cation. The values of $\Delta G_d^\ddagger$ for the complexes studied were within the range 10.3–12.1 kcal mol^{-1}, and it was therefore possible to determine diastereoisomer ratios directly by examining the NMR spectra of the complexes at temperatures where exchange between the diastereoisomers, eg, **28a**$\rightleftharpoons$**28b**, was slow on the NMR time scale [cf. complexes **15** and **16**].

In addition to the measurements of the ratio of the diastereoisomeric complexes the value of $\Delta G_d^\ddagger$ for the process major isomer→minor isomer was obtained from coalescence data. In general, the value of $\Delta G_d^\ddagger$ for a guest *t*-butylammonium cation was less than that for the sterically less demanding

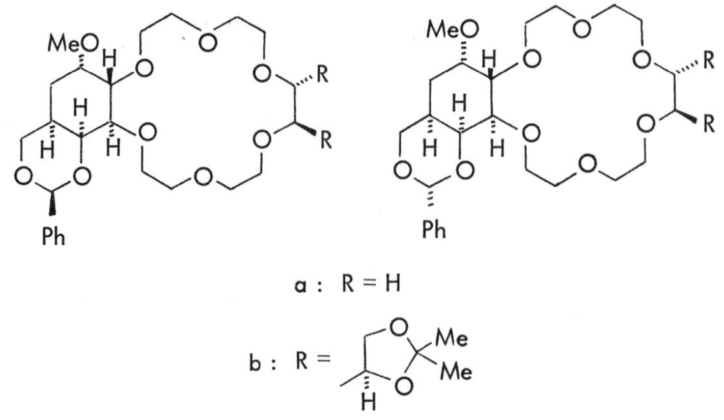

a : R = H

b : R =

26a, b

27a, b

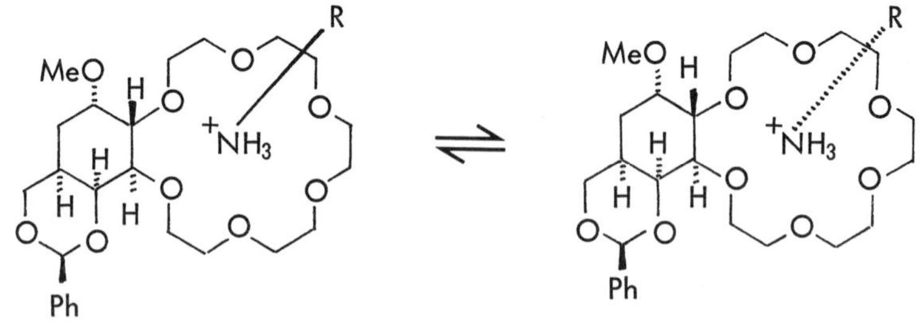

28a

28b

benzyl-ammonium cation or phenylethylammonium cation; this difference was reflected by the lower values of ΔG_a for the *t*-butylammonium cation. In some instances the calculated rates of association of the complexes, obtained from the values of $\Delta G_d^{\ddagger}$ and ΔG_a (see Fig. 1-2a), were characteristic of a diffusion-controlled process, but in other cases they appeared to be significantly slower and to involve a moderately high activation energy. Further studies of this type have been reported by the same group,[44-46] and a number of cases have been noted in which the activation energy for complex formation, $\Delta G_a^{\ddagger}$, is considerably higher than that expected for a diffusion-controlled process. In particular it was shown that in some cases a high value for $\Delta G_a^{\ddagger}$ might be a result of configurational restraints upon the conformation of the substituted 18-crown-6 system, leading to poor co-operativity in the formation of non-covalent bonds between the guest cation and the primary binding site. It was further concluded that this might be an indication of highly directional requirements for the formation of these non-covalent bonds. It appears from these results that the measurement of $\Delta G_d^{\ddagger}$ for any particular guest does not necessarily provide information on the relative magnitudes of ΔG_a for a series of different hosts.

The application of D NMR methods for the study[47-49] of alkylammonium salt complexes of crown dilactones **29** is also of interest, and measured values of

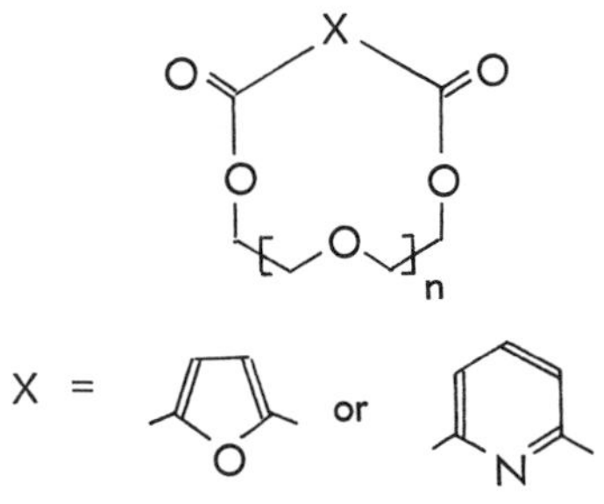

X = or

$n = 3, 4, 5$, etc.

29

$\Delta G_d^{\ddagger}$ are in accord with the known ability of these lactones to form strong complexes with cations. The formation of complexes of $R\overset{+}{N}H_3$ and $\overset{+}{N}H_4$ cations with 18-crown-6 led to an investigation of complex formation between 18-crown-6 and transition-metal ammines by Colquhoun and Stoddart.[50] It was shown that the NMR spectrum of the OCH_2 protons of 18-crown-6 in CD_2Cl_2 showed a change of chemical shift on the addition of 1 mol equiv of $[(Fe(\eta^5\text{-}C_5H_5)(CO)_2(NH_3)]^+[BPh_4]^-$. When the solution was cooled the $—OCH_2—$singlet broadened and gave two signals at $-90°C$ [$T_c - 50°C$], indicating the formation of a face to face complex similar to that formed with alkylammonium cations. A number of such complexes were obtained crystalline, and the crystal structures showed that complex formation was indeed of the type suggested by the NMR study.[51]

The 30-crown-10 derivative **30** forms an interesting crystalline 1:1:1 complex with an (*S*)-$PhCHMe\overset{+}{N}H_3$ cation or a $PhCH_2\overset{+}{N}H_3$ cation and a water molecule.[52] The situation in CD_2Cl_2 solution was investigated by ^1H NMR, and it was shown that the bridgehead protons (see **30**) provided a suitable probe for examination of the face to face exchange of the guest cation and water molecules. Other 1:1:1 complexes were examined in this way, and it was shown that free energies of activation were ca. 10 kcal. mol^{-1} for complexes of the type $\mathbf{30}\cdot H_2O\cdot R\overset{+}{N}H_3$ and $\mathbf{30}\cdot RNH_2\cdot R'\overset{+}{N}H_3$. Structures were not defined fully by the ^1H NMR data, but the observations appeared to be in accord with structures for the complexes in solution similar to that observed in the crystal structure of the complex $\mathbf{30}\cdot H_2O\cdot$(*S*)-$PhCHMe\overset{+}{N}H_3\ C10_4^-$.

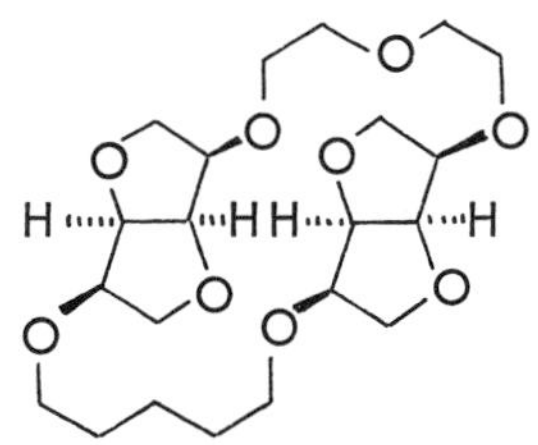

30

The use of NMR spectroscopy to examine the conformational behavior of complexes of monaza crown ethers has been discussed in the previous section. These studies have been extended to a number of diaza crown ethers and to one triaza system. Thus, the diaza-12-crown-4 derivative **31** forms complexes with both secondary and primary alkylammonium cations.[53,54] A recent examination of the complex of **31** with dibenzyl-ammonium thiocyanate by ^{1}H and ^{13}C NMR was interpreted[54] in terms of the interconversion of the two conformations of the complex **32a** and **32b**. These involve the two enantiomeric [3333] conformations of the host macrocycle with the guest cation attached to the same face in both complexes. The free energy of activation associated with the process **32a**⇋**32b** ($\Delta G\ddagger = 7.8$ kcal mol^{-1}) is of the same order as that found for an analogous conformational change in the complexes of the monoaza 15-crown-5 system **8** and presumably involves similar rotation about the OC—CN single bonds and some reorganization of the hydrogen bonding. A further, higher energy barrier associated with dissociation of the complex was found by examining the complex using a 2:1 host–guest ratio of the components. An additional, higher energy barrier ($\Delta G^{\ddagger} = 11$ kcal mol^{-1}) was found for a third process involving face to face exchange of the guest cation and complete conformational inversion of the host macrocycle **33a**⇋**33b**. All three processes

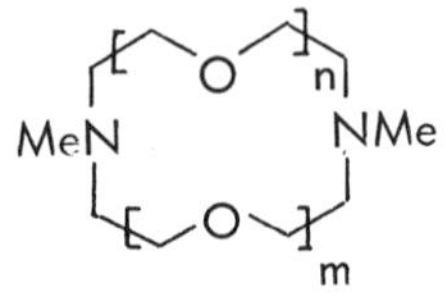

MeN NMe O O n m

31, $m = n = 1$

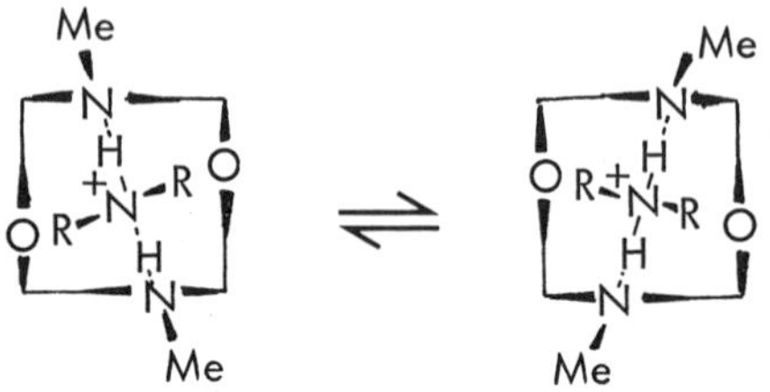

32a ⇌ **32b**

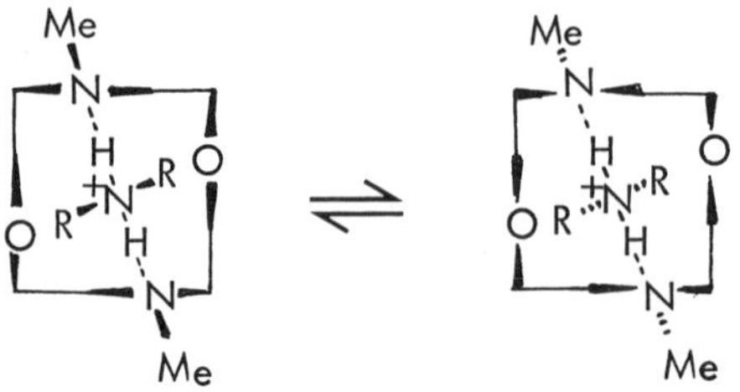

33a ⇌ **33b**

are analogous to those discussed for complexes of the monoaza crown systems **8** and **14**, and it should be noted that this diaza-12-crown-4 system apparently forms only the complex **32** in which the guest has a cis relationship with both *N*-Me groups. Similar energy barriers are also detectable for complexes of the diaza-12-crown-4 system **31** with primary alkylammonium cations.[53]

The diaza-15-crown-5 derivative **34** shows rather similar behavior. The complexes with primary alkylammonium cations also show a preferred cis,cis relationship between the guest cation and the *N*-Me groups as in **35**.[55] Although the exact nature of the complexation is uncertain, it must involve a chiral conformation of the 15-membered ring and hydrogen bonding, probably to the two nitrogen atoms of the macrocycle as shown. The ^{1}H NMR spectrum of the **35**$\cdot PhCH_2\overset{+}{N}H_3NCS^-$ complex in CD_2Cl_2 is well resolved at $-40°C$ and shows evidence for slow face to face exchange of the guest cation (ABCD system for OCH_2CH_2N) with a moderately high energy barrier for the process **35a**$\leftrightharpoons$**35b** ($\Delta G^\ddagger = 13$ kcal mol^{-1}). At very low temperatures ($< -90°C$), changes in the OCH_2CHO signals suggest that a conformational change of the macrocycle becomes slow on the NMR time scale ($\Delta G^\ddagger \approx 8$ kcal mol^{-1}), but the changes are not sufficiently well defined for a precise description of this process. There is no evidence for the presence of any diastereoisomer of the species **35**.

The diaza-18-crown-6 derivatives **36** and **37** form complexes with primary

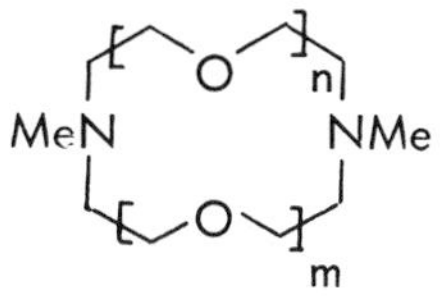

34, $m = 1, n = 2$

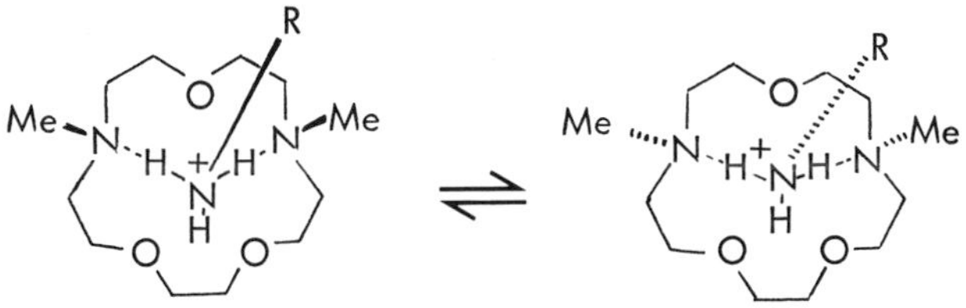

35a $\rightleftharpoons$ **35b**

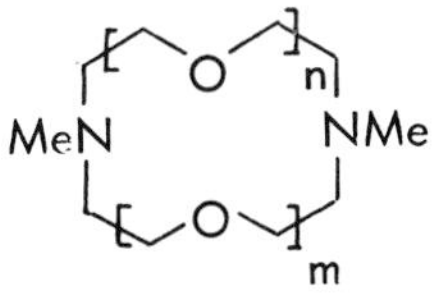

36, $m = 1, n = 3$ 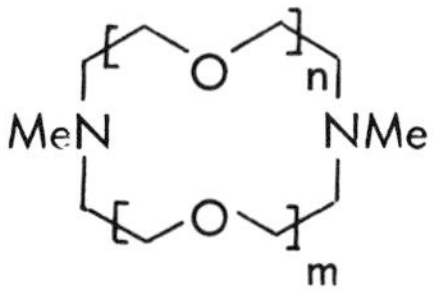 **37**, $m = n = 2$

alkylammonium salts that show very complex temperature dependence of their NMR spectra.[38] At low temperatures both guest and host signals are assignable to the three possible diastereoisomeric complexes **38**, **39** and **40**. This behavior contrasts with the stereoselective complex formation by the hosts **31** and **34**, and it is presumably a consequence of the greater conformational freedom of the larger ring system. It is reasonable to assume that attachment of the $\overset{+}{N}H_3$ group to the macrocycle involves three hydrogen bonds by analogy with the known structure[56] of the complex **5**$\cdot PhCH_2\overset{+}{N}H_3NCS^-$, but the available evidence does not define the structures of the complexes more fully. The dibenzodiaza-18-crown-6 system **41** apparently forms only a single complex with benzyl-ammonium thiocyanate and the ^1H NMR spectrum of the complex is consistent with either a cis,cis (cf. **38**) or trans,trans (cf. **40**) structure. The signals of the guest CH_2 group show a shift to high field, as compared with other complexes, consistent with the trans,trans structure **42**, in which this CH_2 group lies in the shielding zones of the aromatic rings fused to the host macrocycle.

In general, complexation by 21-crown-7 and 24-crown-8 systems involves lower energy barriers for both guest–host exchange (E) and face to face guest exchange ($E+I$) as compared with the smaller ring systems (12-crown-4, 15-crown-5, and 18-crown-6 derivatives). Interesting results for these larger ring systems have been obtained[57] for the 24-crown-6 derivative **43** and the 24-crown-8 derivative **44**. For example, the NMR spectra of 1:1 and 1:2 mixtures of **43** with primary alkylammonium cations in CD_2Cl_2 indicate the formation of a 1:1 complex with a rather low free energy of activation ($\Delta G^{\ddagger}=9$–10 kcal mol^{-1}) for face to face guest exchange. The ^1H NMR signals assignable to the

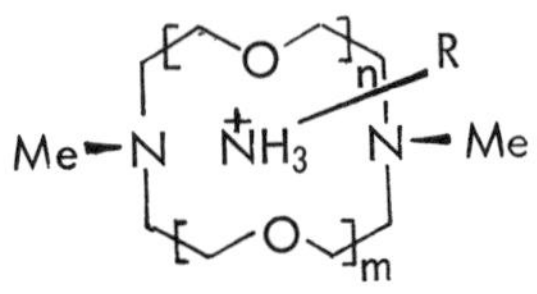

38 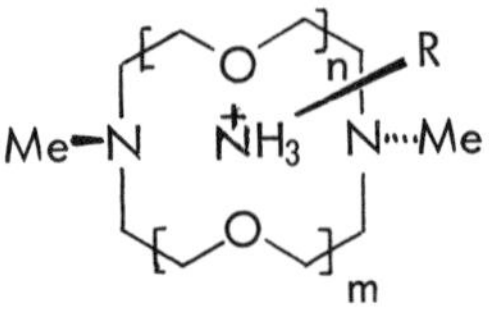 **39** 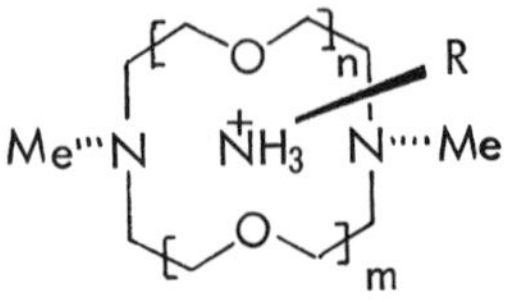 **40**

$m=n=2$ or $m=1$, $n=3$.

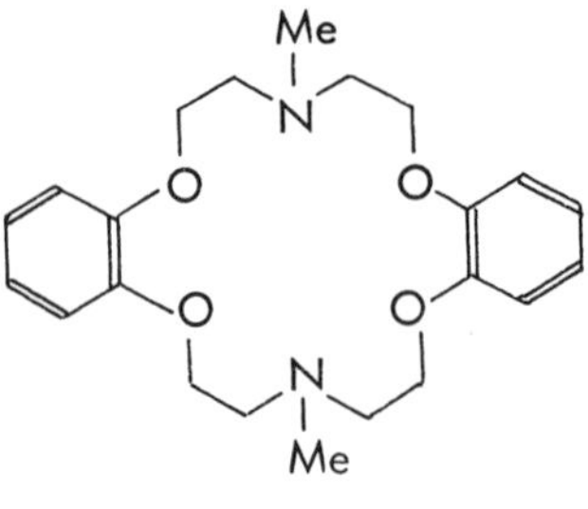

41 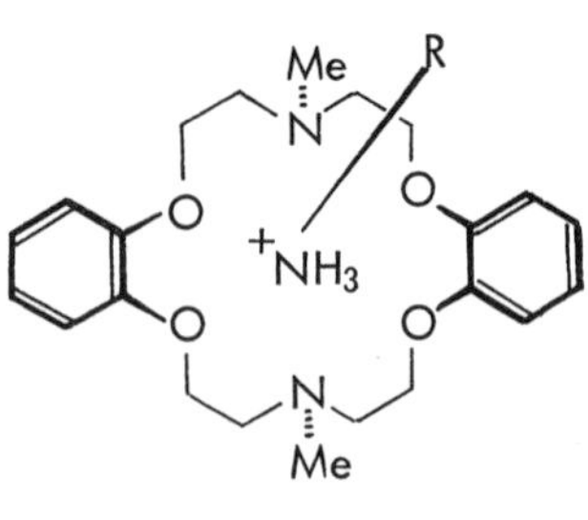 **42**

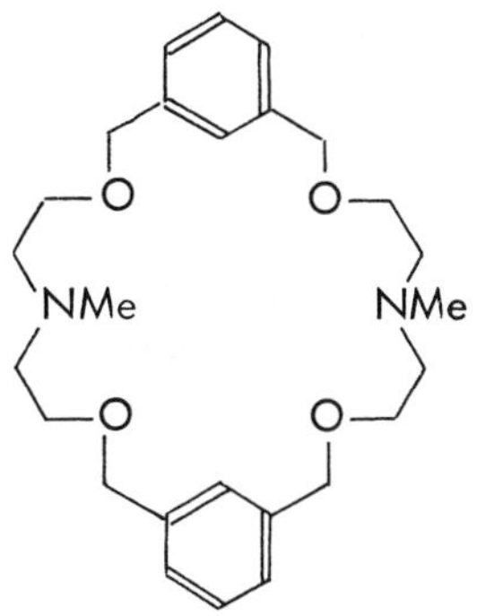

43 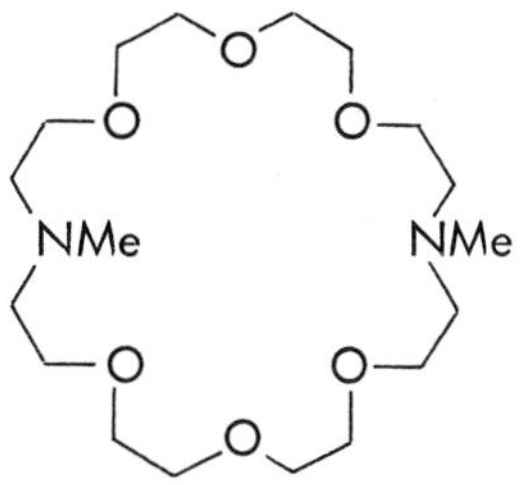 **44**

group R of the guest cation $R\overset{+}{N}H_3$ show considerable shifts to high field as compared with the free cations, and these shifts increase as the temperature is decreased ($\Delta\delta > 2$ ppm for $C\mathit{H}\overset{+}{N}H_3$ and up to 1 ppm for $C\mathit{H}{-}C{-}\overset{+}{N}H_3$ at low temperatures).[58] The NMR data, therefore, suggests that the 1:1 complex adopts the double "nesting" conformation[59,60] **45** in which the guest cation is situated in the shielding zone of the aromatic rings of the host. It is of interest that the host **43** also forms a crystalline 1:2 complex with benzyl-ammonium thiocyanate, which has a structure[57] that is completely different from **45**. In contrast, the 24-crown-8 system **44** in CD_2Cl_2 does not form[57] a well-defined 1:1 complex with benzyl-ammonium thiocyanate, although the 1:2 complex gives a well-resolved ^1H NMR spectrum at $-80°C$ that is consistent with the cis,cis,trans,trans structure **46**.[58] At $-80°C$ the guest $PhC\mathit{H}_2$ gives two distinct signals and the host $NC\mathit{H}_2$ groups are observable as the two H_A triplets and two H_B doublets of *two* different $NCH_AH_BCH_2O$ systems, but the host *N*-Me groups give a single singlet signal over the temperature range $+26°C$ to $-80°C$. At higher temperatures the host NCH_2 signals coalesce to give a single triplet, and the $PhC\mathit{H}_2$ of the guest gives a single singlet. Of the various possible diastereoisomeric forms of a 1:2 complex only the stereochemistry shown in **64** is consistent with all the spectroscopic data. It is of interest that 24-crown-8 is also known to form 2:1 complexes of a rather similar type with metal cations.

There is only one reported example[61] of an examination of complex formation by triaza crown ethers. The triaza-18-crown-6 derivative **47** forms

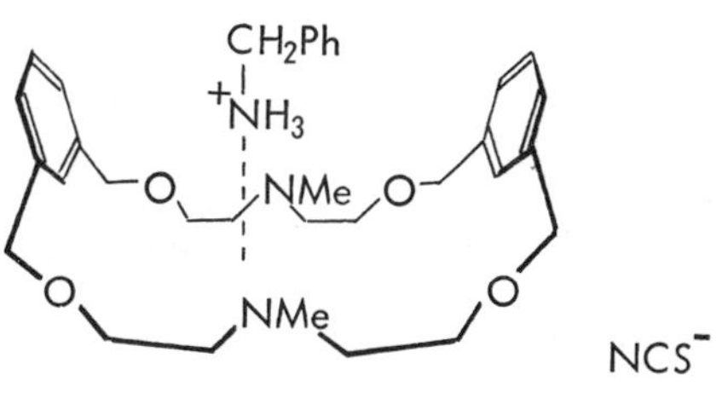

45

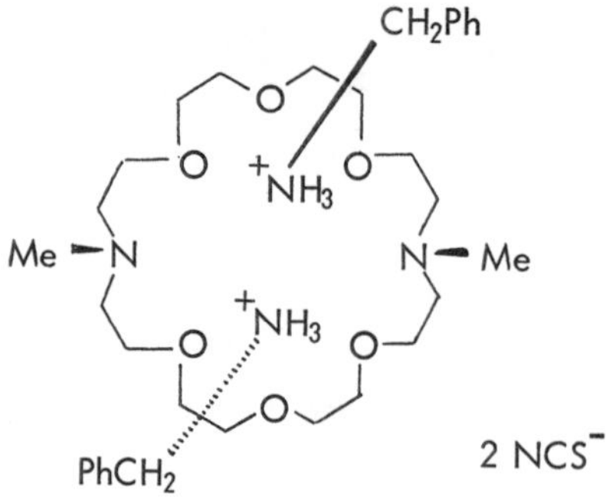

46

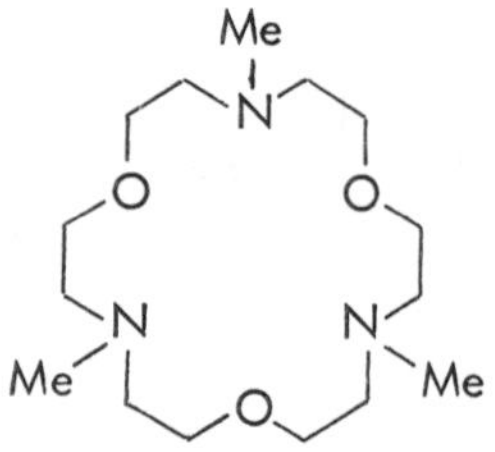

47

very strong complexes with primary alkylammonium cations with values for K_a 20–30 times greater than for 18-crown-6 or diaza-18-crown-6. The NMR spectra of these complexes have not been described in detail, but the NMR spectra of solutions at low temperature indicate the presence of species differing by the orientation of the *N*-Me groups, as expected by analogy with the complexes of the monoaza system **14** and the diaza systems **36** and **37**.

Polycyclic Systems

The guest selectivity of the monocyclic host molecules discussed in the previous section can be enhanced by substitution and the incorporation of steric barriers into the macrocyclic system, and impressive progress has been made with systems of this type.[62,63] An alternative approach is to construct a molecular cavity within the framework of a macropolycyclic system. Examples of such systems, the cryptands **48**, are well known for their selectivity in the complexation of metal cations,[64] and it seemed that a similar approach to the selective complexation of ammonium salts would prove rewarding. The construction of such macropolycyclic systems, eg, **49**, has been described,[65-69] but at the outset of our own work it was clear that the formation of inclusion complexes by hosts

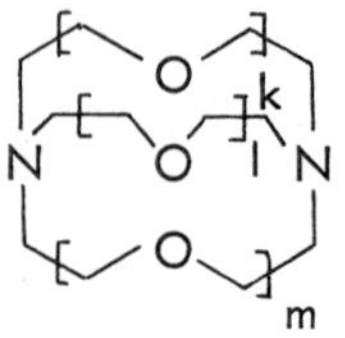

48

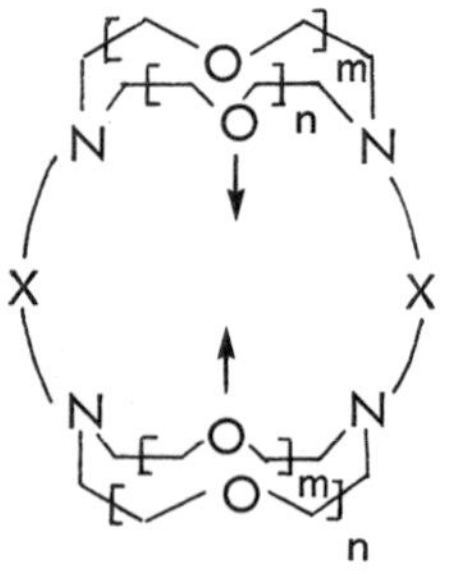

49

such as **49** would require a choice of macrocycles that direct complexation into the cavity, as indicated by the arrows in **49**.

From the preceding section it is clear that the crown rings of the macrotricycles **49** have the required directional properties if they are diaza-12-crown-4 (cf. **32**) or diaza-15-crown-5, (cf. **35**) systems; on the other hand, a diaza-18-crown-6 system would not show the required stereoselectivity (cf. **38**, **39**, and **40**). For maximum guest–host selectivity (recognition) it is important that the cavity in the tricyclic system **49** be well defined, which requires that the bridge contain a rigid group, X. A CH_2—Ar—CH_2 system was chosen to meet this requirement for the bridge because (a) it is structurally rigid and therefore well defined; (b) insertion of such a bridge is a simple synthetic procedure; (c) the aromatic ring in the bridge serves as a probe for the formation of an inclusion complex **50**, as opposed to an addition complex **51**, because the guest in a complex of the type **50** is located in the shielding zone(+)[70] of the aromatic ring (− indicates deshielding); (d) structural variation is simple because a variety of CH_2—Ar—CH_2 systems is available; and (e) the aromatic rings could be used to hold the functional groups required for the development of catalytic systems. A number of tricyclic host molecules **52** were therefore synthesized, and their complexes with alkylammonium cations were examined.[71-75] The results of these studies are discussed separately for systems based upon diaza-12-crown-4 (**52**, $m=n=1$), diaza-15-crown-5 (**52**, $m=1$, $n=2$), and diaza-18-crown-6 (**52**, $m=n=2$).

The simplest host molecule to be examined **52a**[72] has the smallest cavity of the **52** series. The ^{1}H and ^{13}C NMR spectra of the free host molecule provided evidence for the presence in solution of a major and a minor conformation with restricted rotation of the macrocycles and the aromatic rings. For example, at high temperatures the NCH_2CH_2O systems of the diaza-12-crown-4 macrocycles gave rise to an AA′BB′ system that changed to an ABCD system at temperatures below 50°C; similarly, the aromatic ring protons gave a singlet that broadened at low temperatures and was finally resolved as an ABCD system at −100°C. The aromatic ring carbon atoms gave two singlets [2 × (2 C + 4 CH)] at 25°C and signals assignable to two diastereoisomers of different

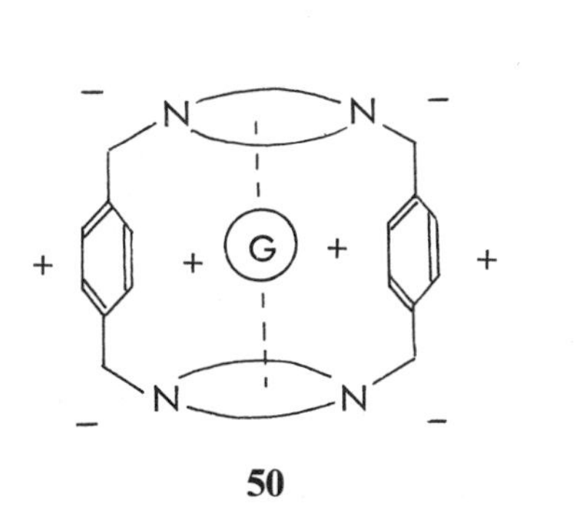

50

51

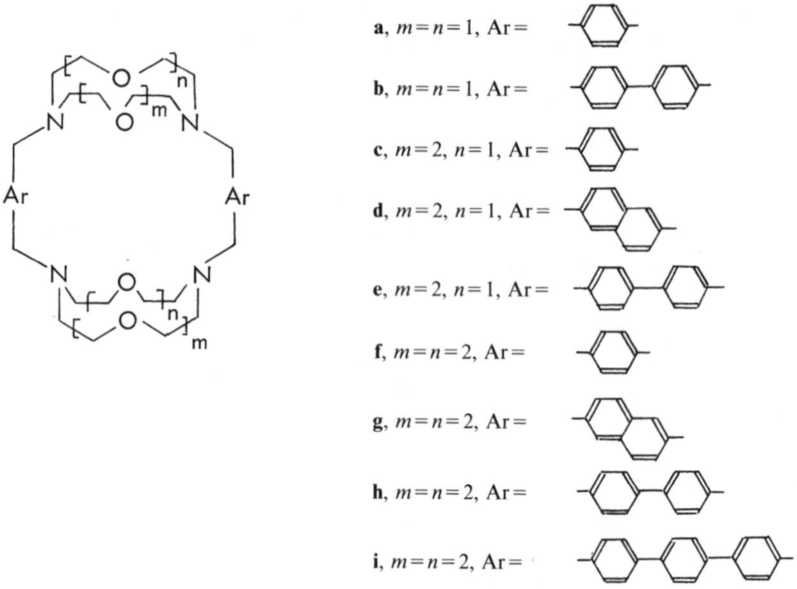

a, $m=n=1$, Ar=

b, $m=n=1$, Ar=

c, $m=2$, $n=1$, Ar=

d, $m=2$, $n=1$, Ar=

e, $m=2$, $n=1$, Ar=

f, $m=n=2$, Ar=

g, $m=n=2$, Ar=

h, $m=n=2$, Ar=

i, $m=n=2$, Ar=

52

symmetry at $-100°C$ [minor species $4\times(C+CH+CH)$ and major species $2\times(C+C+CH+CH+CH+CH)$]. The energy barrier separating these two conformers is low and comparable to that found[53,54] for the interconversion of the [3333] conformations of the diaza-12-crown-4 derivative **31**. The minor conformer could be based upon these [3333] conformations for both 12-membered rings, but it does not appear to be possible to assign a conformation to the major species on this basis.

The ^{1}HMR spectra of 1:1 and 1:2 mixtures of **52a** and $Me\overset{+}{N}H_3NCS^-$ in CD_2Cl_2 provided evidence for the formation of 1:1 and 1:2 complexes. The spectrum of the CH_3 group of the guest cation, in the presence of excess of the host, shows a shift to high field at 25°C (δ 1.17) as compared with the free salt (δ 2.53), consistent with the formation of a 1:1 inclusion complex. The spectrum of a 1:2 mixture of **52a** and $Me\overset{+}{N}H_3NCS^-$ shows a single broad signal (δ 2.07) for the CH_3 group of the guest at 25°C which broadens and collapses at lower temperatures; below $-50°C$ two CH_3 singlets of equal intensity are observable at higher field (δ 0.33) and lower field (δ 3.04) relative to the free salt (δ 2.58) at $-70°C$. The protons of the aromatic rings of the host give a singlet at 25°C and an ABCD system at low temperatures. These results show that the guest cations in the 1:2 complex are situated in two different environments, with only one cation situated in the shielding zone of the aromatic rings. The ^{13}C NMR spectrum of the 1:2 complex can be interpeted in an analogous fashion. It is not possible to define the structure of the 1:2 complex on the basis of these results, but it is possible that one guest cation is situated inside the cavity (cf. **50**) and the second guest cation outside the cavity (cf. **51**), as in structure **53**, rather than

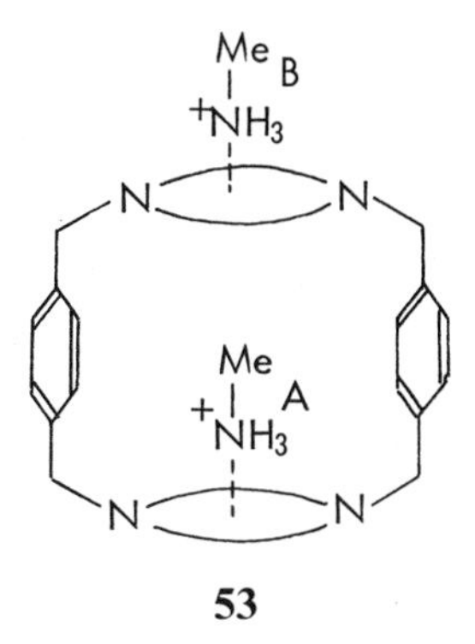

53

in a 1:2 inclusion complex, as in **54**. This conclusion is consistent with the widely different chemical shifts ($\Delta\delta$ 2.71) of the two guest cations and the rather limited space available within the small cavity of the tricyclic host **52a**; it is also consistent with the relatively high energy barrier for the process that interchanges the environments of the guest cations.

The host **52** are ideal receptors for bis-ammonium cations $\overset{+}{N}H_3(CH_2)_n\overset{+}{N}H_3$, and the ^{13}C and ^{1}H NMR spectra of 1:1 mixture of the host **52a** and $\overset{+}{N}H_3(CH_2)_2\overset{+}{N}H_3 \cdot 2\,NCS^-$ in CD_2Cl_2 provide clear evidence for the formation of a strongly bound 1:1 inclusion complex **55**. Thus, at 25°C the ^{1}H NMR spectrum shows a high-field singlet (δ 1.28) for the CH_2 groups of the guest; the shift to high field as compared with the spectrum of the free guest salt (δ 3.10) provides excellent evidence for inclusion of the guest bis-cation as in **55**. A number of changes are observed in the ^{1}H NMR spectrum as the solution is cooled: (a) the protons on the aromatic rings of the host give a singlet at 25°C, which changes to two singlets below -30°C; (b) the OCH_2CH_2N system of the macrocycle gives two broad signals at 25°C, which change to a complex set of multiplets from two ABCD systems at -60°C; and (c) the guest CH_2 signal broadens and collapses and appears as two broad multiplets (probably of an AA′BB′ system, δ_A 2.14, δ_B -0.94) below -70°C. The ^{13}C NMR spectrum of the complex also shows temperature dependence; the aryl-*C*H signals change from a singlet at 25°C to two singlets below -38°C, and this change is accompanied

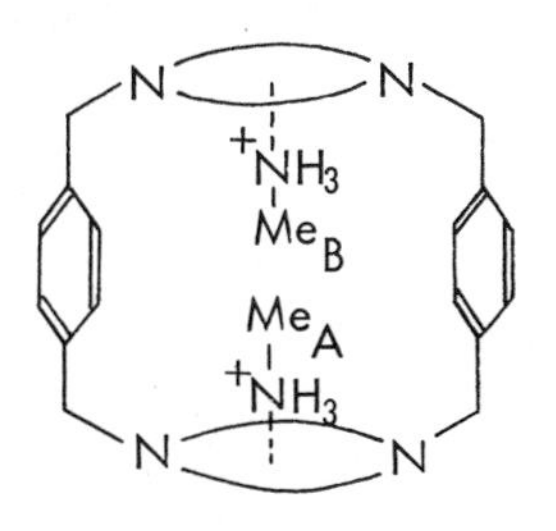

54

55

by similar doubling of the OCH_2 signal, but the guest CH_2 signal remains a singlet over the entire temperature range (+25°C to −83°C).

These changes in the NMR spectra of the C_6H_4 and OCH_2CH_2N units of the host can be associated with hindered rotation of the macrocycles and the aromatic rings in a complex in which the "front" and "back" of the aromatic rings and the macrocycles are in different environments (temperature dependence in the −30°C to −60°C range) and the protons of the CH_2 groups of the guest are diastereotopic (temperature dependence around −70°C). The latter is probably a consequence of a chiral conformation of the macrocycles, for example the [3333] conformation. This information does not define the structure of the complex completely, but it does indicate that there is a unique stereochemistry for complexation in spite of the conformational diastereoisomerism shown by the free host **52a**. The ^1H NMR spectrum of a 2:1 host–guest ratio of **52a** and $\overset{+}{N}H_3(CH_2)_2\overset{+}{N}H_3$ $2NCS^-$ shows separate signals for the aryl protons of the free and complexed host below +5°C indicating a substantial energy barrier ($\approx$13kcal mol^{-1}) for guest exchange.

The cavity of the host **52a**, on the basis of studies of molecular models, is too short to accommodate the next member of the bis-ammonium cation series, $\overset{+}{N}H_3(CH_2)_3\overset{+}{N}H_3$. It is therefore not surprising that the NMR spectra of mixtures of **52a** and the salt $\overset{+}{N}H_3(CH_2)_3\overset{+}{N}H_3 \cdot 2\ NCS^-$ provide only poorly defined evidence for the formation of a complex. Thus, for a 1:1 ratio of the two components the ^1H NMR spectrum shows only small shifts to high field for the guest CH_2 groups at 25°C, although at −50°C broad, high-field signals (δ 1.65 and 1.83) may be assigned to the α-and β-CH_2 groups of the complexed guest cation; however, complexation appears to be incomplete.

The conformational mobility of the aromatic rings and macrocycles of the host **52b** is greater than that of the host **52a**, which has a much smaller cavity. Thus, rotation and inversion of the 12-membered rings only becomes slow below +5°C, and the NCH_2CH_2O system gives a well-resolved ABCD system in the ^1H NMR spectrum of a sample at −40°C. Complexation of methylammonium thiocyanate appears to give, preferentially, a 1:2 complex (host–guest ratio) with identical environments for both guest cations (δ 0.44 at −60°C as compared with δ 2.60 for free $CH_3\overset{+}{N}H_3NCS^-$). If an excess of the guest salt is present, exchange between free and complexed cations becomes slow below −20°C and, at −60°C, separate well-resolved signals are seen for the complexed and free guest salt. There is no evidence for more than one type of 2:1 complex.

Complexation of bis-ammonium salts $\overset{+}{N}H_3(CH_2)_n\overset{+}{N}H_3 \cdot 2\ NCS^-$ by the biphenyl-bridged host **52b** was studied in some detail, because it was expected that optimum complexation would be observed only for salts of the correct length for both terminal $\overset{+}{N}H_3$ groups to be strongly bound by the macrocyclic receptors at the two ends of the host cavity. The ^1H NMR spectra of various mixtures of host **52b** with the salts $\overset{+}{N}H_3(CH_2)_n\overset{+}{N}H_3 \cdot 2\ NCS^-$ provided strong evidence for complex formation with the salts having n=4–7, but the salts $\overset{+}{N}H_3(CH_2)_3\overset{+}{N}H_3 \cdot 2\ NCS^-$ and $\overset{+}{N}H_3(CH_2)_8\overset{+}{N}H_3 \cdot 2\ NCS^-$ gave spectra that provided little or no evidence for complex formation except at very low temperatures. In general, the CH_2 groups of the complexed guest salts showed significant shifts to high field as compared with the uncomplexed salts and exchange between free and complexed salts became slow on the NMR time scale at low temperatures. If an excess of the host tricycle **52b** was present in the solution, it was possible to examine exchange of free and complexed host. The results of these studies are summarized in Table 1-4. The ^1H NMR spectra of the complexed guests generally showed no temperature dependence at temperatures below those at which guest exchange was slow, other than line broadening and chemical shift changes. In all four cases for complexes of the salts $\overset{+}{N}H_3(CH_2)_n\overset{+}{N}H_3 \cdot 2\ NCS^-$, $n=4$–7, changes in the host spectrum at very low temperatures suggested that a conformational change of the macrocycles had become slow on the NMR time scale.

On the basis of the data in Table 1-4 it appears that the complexes of $\overset{+}{N}H_3(CH_2)_5\overset{+}{N}H_3$ and $\overset{+}{N}H_3(CH_2)_6\overset{+}{N}H_3$ are kinetically more strongly bound than those of the other bis-cations. The relative binding energies of the various complexes could be examined more directly by competition experiments. Thus, the NMR spectrum of a solution of the complex **52b**$\cdot\overset{+}{N}H_3(CH_2)_5\overset{+}{N}H_3 \cdot 2$ NCS^- was unchanged by the addition of the salt $\overset{+}{N}H_3(CH_2)_4\overset{+}{N}H_3 \cdot 2\ NCS^-$ except for the appearance of signals assignable to the uncomplexed, shorter bis–cation, showing very clearly that host **52b** shows a strong selectivity for preferential complexation of the larger cation. These competition experiments were most informative at temperatures where cation exchange was slow on the NMR time scale, so that separate signals were observable for free and complexed guest molecules. On the basis of a series of similar experiments for a range of pairs of cations, $\overset{+}{N}H_3(CH_2)_n\overset{+}{N}H_3$ and $\overset{+}{N}H_3(CH_2)_{n+1}\overset{+}{N}H_3$, it was possible to show that the host **52b** selects the cations having $n=5$ and $n=6$ with almost equal preference from the series of cations having $n=3$–8. It is interesting to note that this selectivity for the bis-cations, based upon competition experiments, is in accord with selectivity found by the kinetic method (Table 1-4). This simple picture, however, does not hold for competition between the bis-cations and simple alkylammonium salts, such as $Me\overset{+}{N}H_3 \cdot NCS^-$. Thus, the ^1H NMR spectrum of a 1:2:1 mixture of **52b**, $Me\overset{+}{N}H_3NCS^-$, and $\overset{+}{N}H_3(CH_2)_6\overset{+}{N}H_3 \cdot 2\ NCS^-$ at $-60°C$ is consistent with the formation of almost equal amounts of the complexes **52b**$\cdot 2\ Me\overset{+}{N}H_3$ and **52b**$\cdot\overset{+}{N}H_3(CH_2)_6\overset{+}{N}H_3$ although exchange between free and complexed $Me\overset{+}{N}H_3$ becomes slow below about $-25°C$ and exchange between free and complexed $\overset{+}{N}H_3(CH_2)_6\overset{+}{N}H_3$ becomes slow below about 10°C.

The macrotricycles (**52c–52f**), derived from diaza-15-crown-5, show rather similar properties to the analogous tricycles derived from diaza-12-crown-4. In this case, however, the free macrotricycles can adopt two possible conformations, shown diagramatically in **56a** and **b**, which are related by rotation of one of the crown macrocycles with simultaneous nitrogen inversion. This process fortunately is rapid at room temperature, and it is not necessary to separate diastereoisomeric forms; furthermore, complexation tends to select a single

TABLE 1-4. PROTON NMR SPECTRA (220 MHz) OF BIS-AMMONIUM CATIONS $\overset{+}{N}H_3(CH_2)_n\overset{+}{N}H_3$ IN COMPLEXES WITH HOST **52b** AND TEMPERATURE DEPENDENCE ASSOCIATED WITH GUEST AND HOST EXCHANGE

	^1H NMR spectrum of guest[a]				Host–host exchange[b]		Guest–guest exchange[c]
Guest cation	α-CH_2	β-CH_2	γ-CH_2	δ-CH_2	($\Delta\nu$/Hz	$T_c \pm 5°C$)	($T_c \pm 5°C$)
$\overset{+}{N}H_3(CH_2)_4\overset{+}{N}H_3$	1.05[d]	0.10[d]			77	−40	≈ −50
	0.71[d]	−0.42[d]					
	(2.99)	(1.75)					
$\overset{+}{N}H_3(CH_2)_5\overset{+}{N}H_3$	0.74[e]	0.00[e]	−0.42[e]		77	−5	≈0
	(2.95)	(1.74)	(1.50)				
$\overset{+}{N}H_3(CH_2)_6\overset{+}{N}H_3$	1.05[e]	−0.60[e]	0.07[e]		70	+5	≈ +15
	(2.95)	(1.72)	(1.46)				
$\overset{+}{N}H_3(CH_2)_7\overset{+}{N}H_3$	0.52[f]	0.29[f]	0.04[f]	−0.34[f]	77	−15	≈ −15
	(2.91)	(1.67)	(1.39)	(1.39)			

[a]Chemical shifts in δ for $\overset{+}{N}H_3CH_2(\alpha)$-$CH_2(\beta)$-$CH_2(\gamma)$-, etc. The values in parentheses refer to the uncomplexed bis-cation.
[b]Determined for the aryl proton signals for a solution containing a 1:2 ratio of guest to host.
[c]Determined for the guest CH_2 signals for a solution containing a 2:1 ratio of guest to host. $\Delta\nu$ is very large; hence T_c is only approximate.
[d]At −80°C, at higher temperatures (>0°C) the chemical shifts of the guest showed little evidence of complexation.
[e]At −20°C.
[f]At −60°C.

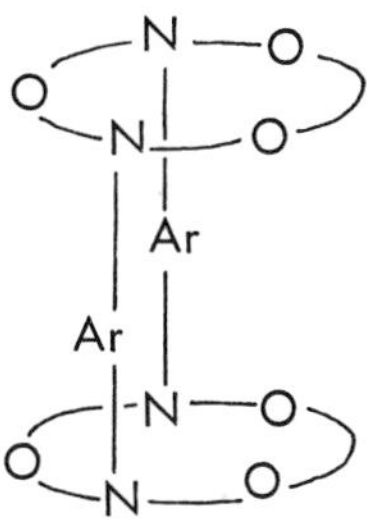

56a 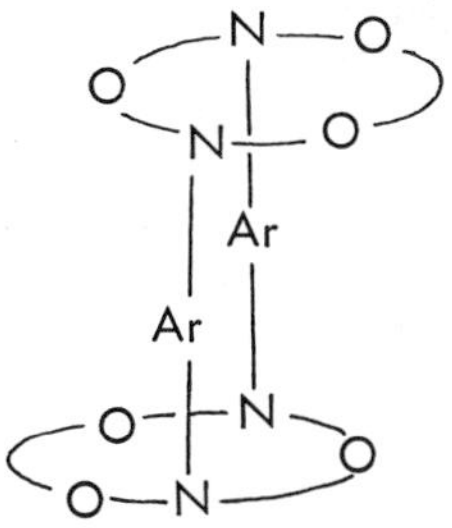 **56b**

conformation from the diastereoisomeric pair **56**. The ^{1}H NMR spectrum of **52c** at +26°C shows two triplet signals for the NCH_2 groups and two triplet signals and a singlet for the OCH_2 groups, assignable to a rapidly interconverting pair of conformations analogous to **56**. At very low temperatures (< −90°C) the OCH_2 and NCH_2 groups are observable as a series of broad signals consistent with a slow rate for the interconversion of these diastereoisomers. The ^{1}H NMR spectrum of a 1:1 mixture of the host **52c** and methylammonium thiocyanate in CD_2Cl_2 at 26°C shows a significant upfield shift for the guest CH_3 group (δ 1.54 as compared with δ 2.54 for the free guest), but the temperature dependence of the spectrum is poorly defined. The spectrum of a 1:2 (host–guest) mixture shows a signal for the guest CH_3 group at intermediate chemical shift (δ 2.02) suggesting the formation of a 1:1 complex. Exchange between free and complexed guest is fast at 26°C and only becomes slow at low temperatures (< −80°C), where the spectrum is poorly defined. The ^{1}H NMR spectra of complexes of host **52c** with bis-ammonium cations $\overset{+}{N}H_3(CH_2)_n\overset{+}{N}H_3$ show the expected high-field shifts for the CH_2 groups of the included guest molecules (Table 1-5) for the bis-cations having $n=2$ to 4. A comparison of the kinetic stabilities of these complexes shows that the bis-cation $\overset{+}{N}H_3(CH_2)_2\overset{+}{N}H_3$ is the preferred guest, and this result is confirmed by competition experiments. The structures of the complexes are not totally defined by the NMR data, but the spectra are consistent with the formation of a single 1:1 complex in each case in spite of the potential diastereoisomerism of the host (cf. **56a** and **b**).

The host **52d** has a larger cavity than **52c** and potentially greater tendency for conformational isomerism because of the lower symmetry of a 2,6-disubstituted naphthalene system as compared with a 1,4-disubstituted benzene ring. Therefore, **52d** can exist in four diastereoisomeric conformations, **57a–d** related by rotation of the 15-membered aza crown macrocycles or by rotation of the napthalene system. This provides five different environments, A–E (see **57**) for a complexed monocation. This multiplicity of environment is reflected in the ^{1}H NMR spectrum of the 1:2 complex **52d**·2 ($Me\overset{+}{N}H_3 \cdot NCS^-$), which shows four of the five possible guest CH_3 signals at −83°C. At higher temperatures conformational interconversion becomes fast on the NMR time scale, to give two broad signals at −36°C and finally a singlet at +37°C. Although it is not

TABLE 1-5. PROTON NMR SPECTRA[a] (220 AND 400 MHz) OF BIS-AMMONIUM CATIONS $\overset{+}{N}H_3(CH_2)_n\overset{+}{N}H_3$ IN COMPLEXES WITH HOSTS **52c–e**

Host	Guest cation, $\overset{+}{N}H_3(CH_2)_n\overset{+}{N}H_3$, n	Temp. (°C)	^{1}H NMR spectrum of guest			
			α-CH_2	β-CH_2	γ-CH_2	δ-CH_2
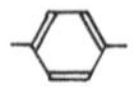 **52c**	2	35	1.55			
		−110	0.37, 1.70			
	3	−60	0.96	1.30		
	4	−60	1.60	0.30		
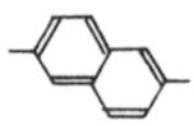 **52d**	4	37	0.94	−0.37		
		−36	0.41, 0.72	−1.03, −0.32		
		−104	{1.12, 0.81 {0.02, −0.06	{−0.27, 0.61 {−0.85, −1.78		
	5	−36	1.29, 0.63	0.18, −0.32	−0.49	
	6	−9	0.98	0.05	−0.55	
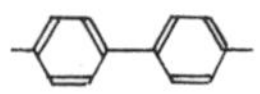 **52e**	4	26	2.64	1.51		
	5	−50	1.00	0.60	−0.25	
	6	−40	1.10	0.20	0.20	
	7	+45	2.00	0.90	0.80	0.80

[a]For solutions in CD_2Cl_2 containing a small proportion of CD_3OD at the stated temperatures.

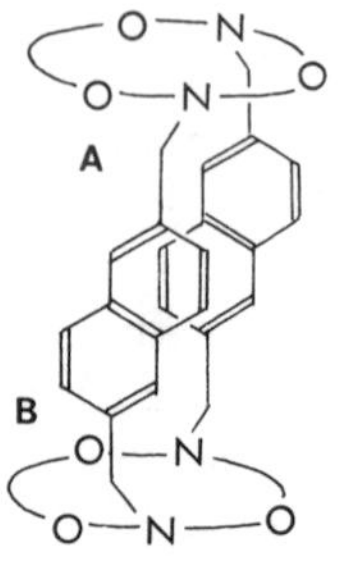

57a 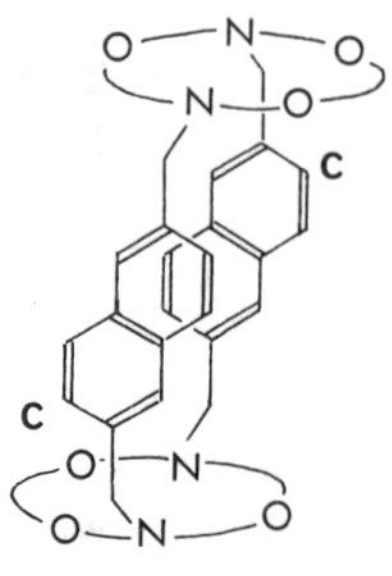 **57b** 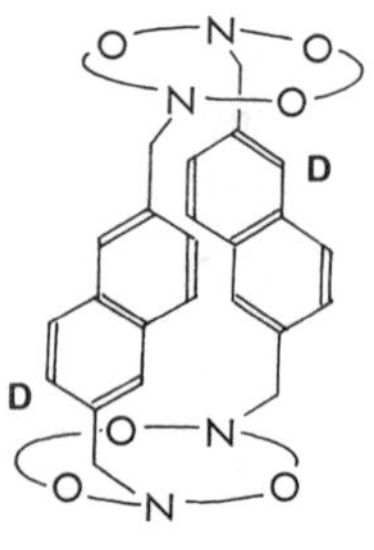 **57c** 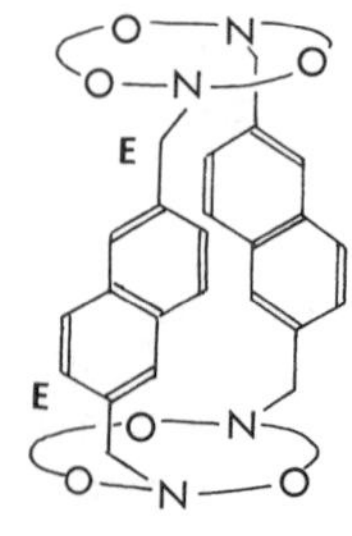 **57d**

possible to assign individual CH_3 signals to the different 1:2 complexes based upon the four conformations **57a–d**, a minimum of three of these conformations must be involved in complexation. It is interesting to note that even for a 1:1 ratio of **52d** and $Me\overset{+}{N}H_3 \cdot NCS^-$ the 1:1 complex is formed in only a minor amount, and the majority of the complexation involves the formation of the same mixture of diastereoisomeric 1:2 complexes.

The complexation of bis-ammonium cations $\overset{+}{N}H_3(CH_2)_n\overset{+}{N}H_3$ by host **52d** shows much greater stereoselectivity. Thus, the ^{1}H NMR spectrum of the 1:1 complex **52d**$\cdot\overset{+}{N}H_3(CH_2)_4\overset{+}{N}H_3$ shows two guest signals at +37°C corresponding to the α-and β-CH_2 groups with the expected shifts to high field. At +26°C these signals are very broad, and at −32°C the spectrum shows four CH_2 signals for the guest in a major species of complex and two additional, low-intensity signals for a minor species. Finally, at very low temperatures the four signals of the major species broaden and eventually give eight signals in the range δ 1 to −2 below −90°C, indicating a different environment for *each* of the eight guest methylene protons. This remarkable spectroscopic behavior is consistent with a major 1:1 complex formed by the conformation **57a**, in which the two ends of the guest salt occupy different environments; at high temperatures the guest is able to rotate within the cavity or apparently rotates as a result of dissociation and re-association. At very low temperatures the protons of each CH_2 group become visibly diastereotopic, presumably because the interconversion of enantiomeric chiral conformations of the 15-membered rings becomes slow on the NMR time scale. The aromatic protons of the host are observable as two ABCD systems at −36°C and finally as four rather broad, overlapping ABCD systems below −90°C. Thus the complex must have C_1 symmetry, and at very low temperatures this low symmetry is reflected in the signals of both guest and host. In accord with this interpretation the ^{13}C NMR spectrum of the guest is observable as two signals at 30°C which broaden and separate to give four signals at −60°C. The ^1H NMR spectrum of the complex **52d**·$\overset{+}{N}H_3(CH_2)_5\overset{+}{N}H_3$ shows rather similar temperature dependence, although the spectrum of the guest at very low temperatures is less well resolved. On the other hand, the spectra of the complexes of $\overset{+}{N}H_3(CH_2)_3\overset{+}{N}H_3 \cdot 2\ NCS^-$ and $\overset{+}{N}H_3(CH_2)_6\overset{+}{N}H_3 \cdot 2\ NCS^-$ are less well defined, do not show clear evidence for the non-equivalence of the two ends of the guest bis-cation, and are weaker complexes on a kinetic basis. This conclusion, based upon the temperature dependence of the spectra, is confirmed by competition experiments. These show (a) no evidence for complexation of the bis-cation $\overset{+}{N}H_3(CH_2)_3\overset{+}{N}H_3$ in the presence of the larger cation $\overset{+}{N}H_3(CH_2)_4\overset{+}{N}H_3$, (b) no evidence for complexation of the cation $\overset{+}{N}H_3(CH_2)_6\overset{+}{N}H_3$ in the presence of the shorter cation $\overset{+}{N}H_3(CH_2)_5\overset{+}{N}H_3$, and (c) a preference of about 7:1 for complexation of the guest $\overset{+}{N}H_3(CH_2)_4\overset{+}{N}H_3$ rather than $\overset{+}{N}H_3(CH_2)_5\overset{+}{N}H_3$. This host, therefore, shows remarkably high recognition for the guest bis-cations.

The host **52e** has the longest cavity of the compounds examined in the tricyclic series based upon diaza 15-crown-5. It forms 1:1 inclusion complexes with the salts $\overset{+}{N}H_3(CH_2)_n\overset{+}{N}H_3 \cdot 2\ NCS^-$, $n = 5$–7 and, on the basis of competition experiments, shows a strong preference for the bis-cation $\overset{+}{N}H_3(CH_2)_6\overset{+}{N}H_3$. These three hosts **52c–52e** all show excellent evidence in their NMR spectra for the formation of 1:1 inclusion complexes; the details of the ^1H NMR spectra of the guest molecules and some of the temperature dependence of the spectra associated with guest–host exchange and other processes are summarized in Table 1-5. The results of competition experiments for the hosts **52a–e** and the guest bis-cations $\overset{+}{N}H_3(CH_2)_n\overset{+}{N}H_3$ are summarized in Table 1-6.

Comparison of the length of the cavity (l), as defined by the N–CH_2ArCH_2–N distance, and the $\overset{+}{N}$--$\overset{+}{N}$ separation (d) in the extended conformation of the bis-cation shows that complexation is optimized for the

TABLE 1-6. SELECTIVITY FOR GUESTS BIS-CATIONS $\overset{+}{N}H_3(CH_2)_n\overset{+}{N}H_3$ IN COMPLEXATION BY HOSTS **52 a–e**

Host	Selectivity[a]	Length of cavity (l) (Å)	Length of preferred guest (d) (Å)
52a, Ar = 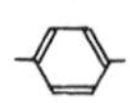	$n=2 \gg 3$	5.8	3.6
52b, Ar = 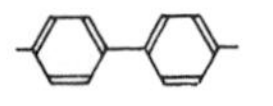	$n=3 \ll 4 < 5 = 6 > 7 \gg 8$	10.1	7.9[b]
52c, Ar = 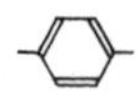	$n=2 > 3 > 4$	5.8	3.6
52d, Ar = 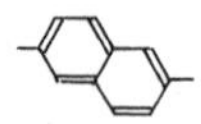	$n=3 \ll 4 > 5 > 6$	7.9	6.1
52e, Ar = 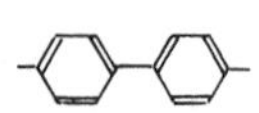	$n=4 < 5 < 6 > 7$	10.1	8.5

[a]For solutions in CD_2Cl_2 containing a small proportion of CD_3OD Selectivity for $\overset{+}{N}H_3(CH_2)_n\overset{+}{N}H_3$ indicated in the table by the appropriate value n and inequality or equality signs.
[b]Average of bis-cations with $n=5$ and $n=6$.

relationship $l = d - 2$ Å. This relationship is consistent with complexation analogous to that shown in **55**.

Polycyclic cryptands have also been investigated by Lehn and co-workers[74–76] as receptors for ammonium and bis-ammonium cations. Thus, the tricyclic diaza-18-crown-6 derivatives **52f–52i** form complexes with the bis-ammonium cations $\overset{+}{N}H_3(CH_2)_n\overset{+}{N}H_3$ in which the guest salt is enclosed within the cavity of the tricyclic system. The ^1H NMR spectra of these complexes show substantial upfield shifts (up to 2–3 ppm) for the guest CH_2 groups, as compared with the same guest bound to two diaza-18-crown-6 macrocycles. These upfield shifts correlate well with the results of an investigation of the dynamic properties of the complexes using ^{13}C relaxation times. Molecular motion within a complex is coupled to an extent that depends upon the degree of structural complementarity between the two components.[76] The extent of this coupling may be investigated by measuring the ^{13}C relaxation times T_1 for the two components of the complex. For a $^{13}CH_n$ group the ^{13}C correlation time is given by

$$T_1^{-1} = 3.60 \times 10^{10} n r_{C-H}^{-6} \tau_C \qquad (1\text{-}1)$$

where r_{C-H} is the C–H separation in Å (1.10 ± 0.02 Å). The dynamic coupling coefficient (χ) is the ratio of substrate correlation time to ligand correlation time. The latter is defined for the hosts **52** as the correlation time for the Ar*C*H_2N ^{13}C nucleus because this is generally the host carbon atom with the longest correlation time. The value of χ is a good indication of the extent to which motions of the host and guest components are coupled and hence of the closeness of fit between the two components.[75,76] Complexation of the hosts **52** slows down molecular motion by a factor of 2 or 3 and, based upon the main features of τ_C and χ values (Table 1-7), three types of behavior are recognized:

TABLE 1-7. CORRELATION TIMES τ_c AND DYNAMIC COUPLING COEFFICIENTS χ OBTAINED FROM ^{13}C RELAXATION TIMES FOR COMPLEXES OF **52 f–i** WITH BIS-AMMONIUM CATIONS $\overset{+}{N}H_3(CH_2)_n\overset{+}{N}H_3$

Host	Guest $\overset{+}{N}H_3(CH_2)_n\overset{+}{N}H_3$	CH_2N	τ_c values[a]					
			α-CH_2	β-CH_2	γ-CH_2	δ-CH_2	ε-CH_2	ζ-CH_2
52f	—	60						
	3	110	105 (0.95)	105 (0.95)				
52g	—	60						
	5	155	170 (1)	170 (1)	160 (0.94)			
	6	160	135 (0.77)	125 (0.71)	120 (0.69)			
	7	150	90 (0.60)	80 (0.53)	90 (0.60)	100 (0.67)		
52h	—	75						
	7	170	155 (0.91)	135 (0.79)	150 (0.88)	140 (0.82)		
	8	150	120 (0.77)	100 (0.65)	125 (0.81)	110 (0.71)		
	9	230	90 (0.39)	110 (0.48)	95 (0.41)	70 (0.30)	80 (0.35)	
52i	—	100						
	8	325	125 (0.47)	90 (0.34)	100 (0.38)	85 (0.32)		
	9	200	155 (0.77)	90 (0.45)	85 (0.42)	105 (0.52)	105 (0.52)	
	10	190	155 (0.77)	135 (0.67)	135 (0.67)	160 (0.80)	150 (0.75)	
	11	200	160 (0.80)	100 (0.50)	90 (0.45)	90 (0.45)	90 (0.45)	110 (0.55)
	12	200	155 (0.77)	85 (0.42)	85 (0.42)	85 (0.42)	115 (0.57)	90 (0.45)

[a] For solutions in $CDCl_3$–CD_3OD (9:1) at 35°C using guest dipicrate salts, τ_c in picoseconds.

(a) strong dynamic coupling (χ closest to 1) is observed for the complexes **52f**·$\overset{+}{N}H_3(CH_2)_3\overset{+}{N}H_3$, **52g**·$\overset{+}{N}H_3(CH_2)_5\overset{+}{N}H_3$, **52h**·$\overset{+}{N}H_3(CH_2)_7\overset{+}{N}H_3$, and **52i**·$\overset{+}{N}H_3(CH_2)_{10}\overset{+}{N}H_3$, indicating that substrate host complementarity is maximized for these cases (Table 1-6). As the length of the optimal substrate chain increases, χ decreases, indicating that chains of the bound substrate are, as expected, more flexible within the larger cavities; the authors noted a corresponding decrease in high-field shifts of the guest CH_2 protons in these cases.

(b) When the substrate is too long or too short to fit the cavity, χ decreases, largely as a result of increasing mobility (shorter τ_c) of the $(CH_2)_n$ chain of the guest.

(c) For cases in which the substrate shows virtually no high-field shift for the CH_2 protons [(**52h**·$\overset{+}{N}H_3(CH_2)_9\overset{+}{N}H_3$ and **52i**·$\overset{+}{N}H_3(CH_2)_8\overset{+}{N}H_3$], motion of the host is significantly slower and that of the guest significantly faster than for complexes in which the two components are more closely matched. This suggests that ion pairing or aggregation may occur, rather than the formation of an inclusion complex analogous to **50**. This examination of the dynamics of molecular complexes illustrates an important alternative procedure for examining molecular complexes of the inclusion type.

The tetracyclic hosts **58** have also been examined by Lehn's group.[77] These hosts formed complexes with the bis-ammonium picrates $\overset{+}{N}H_3(CH_2)_5\overset{+}{N}H_3$ and $\overset{+}{N}H_3(CH_2)_6\overset{+}{N}H_3$ in which the CH_2 groups of the guest cations showed the

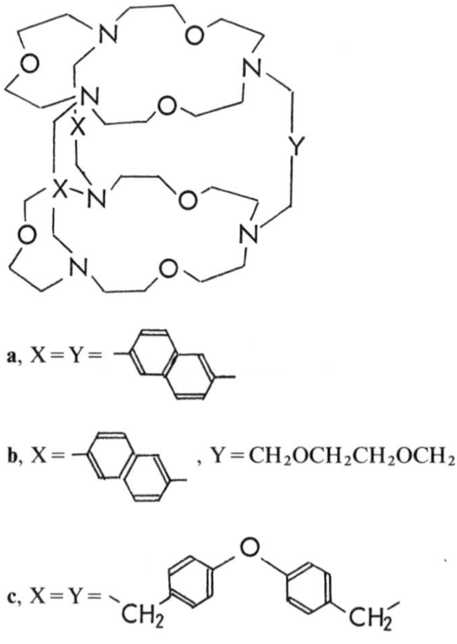

a, X=Y=

b, X= , Y=$CH_2OCH_2CH_2OCH_2$

c, X=Y=

58

R=CH_3, X=

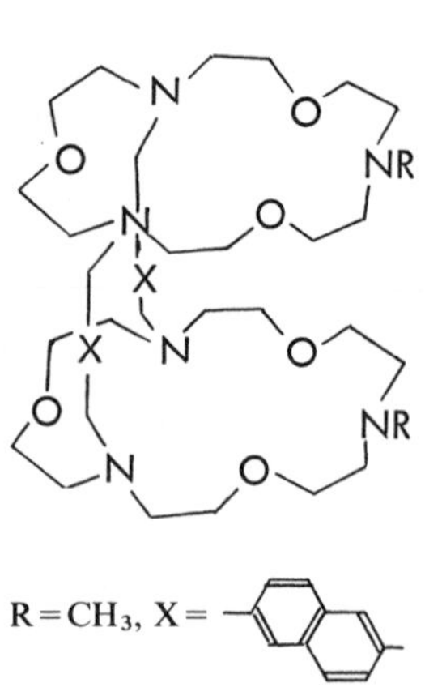

59

TABLE 1-8. PROTON NMR SPECTRA[a] OF BIS-AMMONIUM CATIONS IN COMPLEXES WITH HOSTS **58a** AND **b**, **59**, AND **52g**

	$\overset{+}{N}H_3(CH_2)_5\overset{+}{N}H_3$			$\overset{+}{N}H_3(CH_2)_6\overset{+}{N}H_3$		
Host	α-CH_2	β-CH_2	γ-CH_2	α-CH_2	β-CH_2	γ-CH_2
58a, Ar= 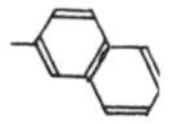	1.47	−1.35	−1.13	1.51	−0.76	−1.85
58b, Ar= 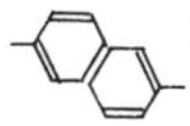	1.62	−0.23	−0.90			
59, Ar= 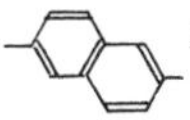	2.06	0.67	−0.52			
52g, Ar= 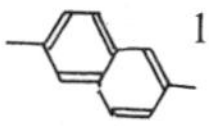	1.95	−0.22	−1.10	2.5	0.31	−0.28

[a]At 18°C for solutions in $CDCl_3$–CD_3OD (9:1).

expected shifts to high field in their ^1H NMR spectra, indicating the formation of inclusion complexes. Considerable discrimination between guest cations was found, as for the tricyclic systems **52**; in particular, **58a** does not form complexes with the cations $\overset{+}{N}H_3(CH_2)_4\overset{+}{N}H_3(CH_2)_8\overset{+}{N}H_3$. Competition experiments showed that the complexes **58a**·$\overset{+}{N}H_3(CH_2)_5\overset{+}{N}H_3$ and **52g**·$\overset{+}{N}H_3(CH_2)_5\overset{+}{N}H_3$ are of similar stability and that the tighter binding of substrate by the triaza-18-crown-6 macrocycles of **58a** is offset by increased interaction between the substrate and the three bridges of the macrotetracyclic system as compared with the two bridges of the macrotricyclic system **52g**. Details of chemical shifts of guest dications in complexes of **58a** and **b**, **59**, and **52g** are given in Table 1-8.

Conclusions

The study of crown ether ammonium cation complexes by D NMR spectroscopy has provided information on the dynamic behavior of the complexes in solution. In particular, it has been possible to detect processes that involve guest–host exchange, probably through intermediate uncomplexed guest and host species, and conformational changes that occur within the complex with no requirement for dissociation. Detection of such processes appears to be rather generally feasible for complexes of primary alkylammonium salts with 12-, 15-, and 18-membered crown and aza crown host macrocycles; it also may be possible in some cases for hosts based upon larger macrocycles. For some hosts, complexation is stereoselective and leads to a single species of complex, but in other cases a mixture of diastereoisomeric complexes may be formed. In particular, a cis relationship is found between the side chains on the nitrogen

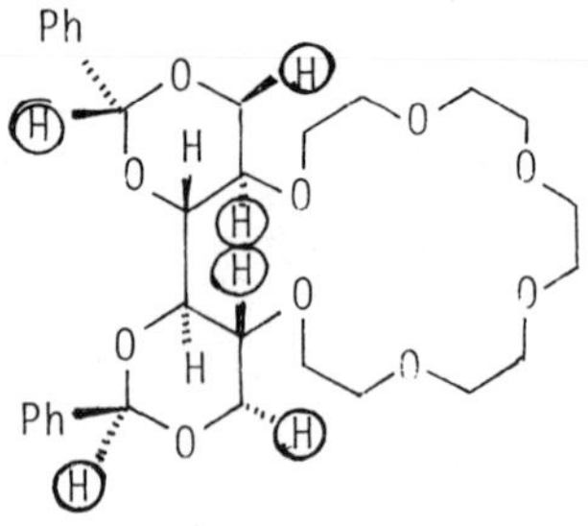

60

atoms of 12- and 15-membered aza crown systems and the guest alkylammonium cation.

On the basis of these studies of monocyclic systems, it has been possible to design polycyclic aza crown hosts that form inclusion complexes with both simple alkylammonium cations and with bis-ammonium cations. These inclusion complexes may be formed with very high guest–host selectivity and stereoselectivity if the polycyclic system includes rigid structural units. Inclusion complexes may be examined by D NMR lineshape methods, but, in addition, measurement of dynamic coupling between guest and host using ^{13}C T_1 measurements may provide valuable information on the closeness of guest–host fit.

Finally, it is interesting to note that *intermolecular* NOEs, previously recognized for complexes of the vancomycin group of antibiotics, have been reported[78] recently for the benzylammonium perchlorate complex of the crown ether analog **60** (from the C_6H_5 protons of the guest to the encircled protons and to the OCH_2CH_2O protons of host **60**). This study also noted saturation transfer between protons of the host related by exchange processes. The wider use of high-field Fourier transform NMR spectrometers for studies of this type should provide increasingly detailed evidence for the structures of complexes in solution, which is complementary to that obtained for the solid phase by the more conventional use of x-ray crystallography.

References

1. Stoddart, J. F. In "Comprehensive Organic Chemistry", Barton, D. H. R.; Ollis, W. D., Eds.; Pergamon: Oxford, 1979; Vol. 1, p. 3.
2. Barton, D. H. R. *Experientia* **1950**, *6*, 316.
3. Lehn, J. M. *Pure Appl. Chem.* **1978**, *50*, 871; **1979**, *51*, 979.
4. Bradshaw, J. S.; Maas, G. E.; Izatt, R. M.; Christensen, J. J. *Chem. Rev.* **1979**, *79*, 37.
5. Izatt, R. M.; Christensen, J. J. "Synthetic Multidentate Macrocyclic Compounds", Academic: New York, 1978.
6. DeJong, F.; Reinhoudt, D. N. *Adv. Phys. Org. Chem.* **1980**, *17*, 279.
7. Reinhoudt, D. N.; DeJong, F. In "Progress in Macrocyclic Chemistry", Izatt, R. M.; Christensen, J. J., Eds.; Wiley: New York, 1979; p. 157.
8. Bender, M. L.; Komiyama, M. "Cyclodextrin Chemistry"; Springer Verlag: New York, 1977.

9. Bender, M. L.; Komiyama, M. "Cyclodextrin Chemistry"; Springer Verlag: New York, 1977.
10. Tabushi, I.; Nabeshima, T.; Kitaguchi, H.; Yamamura, K. *J. Am. Chem. Soc.* **1982**, *104*, 2017.
11. Tabushi, I.; Kimura, Y.; Yamamura, K. *J. Am. Chem. Soc.* **1978**, *100*, 1304; **1981**, *103*, 6486.
12. Odashima, K.; Itai, A.; Iitaka, Y.; Koga, K. *J. Am. Chem. Soc.* **1980**, *102*, 2504.
13. Odashima, K.; Soga, T.; Koga, K. *Tetrahedron Lett.* **1981**, 5311.
14. Dobler, M. "Ionophores and their Structures"; Wiley: New York, 1981.
15. Bongini, A.; Feeney, J.; Williamson, M. P.; Williams, D. H. *J. Chem. Soc., Perkin Trans. 2* **1981**, 201.
16. Williamson, M. P.; Williams, D. H. *J. Am. Chem. Soc.* **1981**, *103*, 6580.
17. Williams, D. H.; Butcher, D. W. *J. Am. Chem. Soc.* **1981**, *103*, 5697.
18. Harris, C. M.; Harris, T. M. *J. Am. Chem. Soc.* **1982**, *104*, 4293.
19. Harris, C. M.; Harris, T. M. *J. Am. Chem. Soc.* **1982**, *104*, 363.
20. Kalman, J. R.; Williams, D. H. *J. Am. Chem. Soc.* **1980**, *102*, 906.
21. McGahren, W. J.; Martin, J. H.; Morton, G. O.; Hargreaves, R. T.; Leese, R. A.; Lovell, F. M.; Ellestad, G. A.; O'Brien, E.; Holker, J. S. E. *J. Am. Chem. Soc.* **1980**, *102*, 1671.
22. Ellestad, G. A.; Leese, R. A.; Morton, G. O.; Barbatschi, F.; Gore, W. E.; McGahren, W. J.; Armitage, I. M. *J. Am. Chem. Soc.* **1981**, *103*, 6522.
23. Sheldrick, G. M.; Jones, P. G.; Kennard, O.; Williams, D. H.; Smith, G. A. *Nature* (London) **1978**, *271*, 223.
24. Pedersen, C. J. *J. Am. Chem. Soc.* **1967**, *89*, 2495, 7017; 1970, *92*, 386, 391; *J. Org. Chem.* **1971**, *36*, 254, 1690.
25. Pedersen, C. J.; Frensdorff, H. K. *Angew. Chem. Int. Ed. Engl.* **1972**, *11*, 16.
26. Liesegang, G. W.; Eyring, E. M. In "Synthetic Multidentate Macrocyclic Compounds", Izatt, R. M.; Christensen, J. J., Eds.; Academic: New York, 1978; p. 254.
27. Kyba, E. B.; Koga, K.; Sousa, L. R.; Siegel, M. G.; Cram, D. J.; *J. Am. Chem. Soc.* **1973**, *95*, 2692.
28. Helgeson, R. C.; Timko, J. M.; Moreau, P.; Peacock, S. C.; Cram, D. J. *J. Am. Chem. Soc.* **1974**, *96*, 6762.
29. Timko, J. M.; Moore, S. S.; Walba, D. M.; Hiberty, P. C.; Cram, D. J. *J. Am. Chem. Soc.* **1977**, *99*, 4207.
30. Leigh, S. J.; Sutherland, I. O. *J. Chem. Soc. Chem. Commun.* **1975**, 414.
31. Binsch, G. *Top. Stereochem.* **1968**, *3*, 97.
32. Sutherland, I. O. *Annu. Rept. NMR Spectrosc.* **1971**, *4*, 71.
33. Lincoln, S. F. *Prog. React. Kinet.* **1977**, *9*, 1.
34. Hodgkinson, L. C.; Leigh, S. J.; Sutherland, I. O. *J. Chem. Soc., Chem. Commun.* **1976**, 639, 640.
35. Johnson, M. R.; Sutherland, I. O.; Newton, R. F. *J. Chem. Soc., Perkin Trans. 1* **1979**, 357.
36. Leigh, S. J.; Sutherland, I. O. *J. Chem. Soc., Perkin Trans. 1* **1979**, 1089.
37. Hodgkinson, L. C.; Sutherland, I. O. *J. Chem. Soc., Perkin Trans. 1* **1979**, 1908.
38. Hodgkinson, L. C.; Johnson, M. R.; Leigh, S. J.; Spencer, N.; Sutherland, I. O.; Newton, R. F. *J. Chem. Soc., Perkin Trans. 1* **1979**, 2193.
39. DeJong, F.; Reinhoudt, D. N.; Smit, C. J.; Huis, R. *Tetrahedron Lett.* **1976**, 4783.
40. DeJong, F.; Reinhoudt, D. N.; Huis, R. *Tetrahedron Lett.* **1977**, 3985.
41. DeJong, F.; Reinhoudt, D. N.; Torny, G. J. *Tetrahedron Lett.* **1979**, 911.
42. Reinhoudt, D. N.; Den Hertog, H. J. Jr; DeJong, F. *Tetrahedron Lett.* **1981**, 2513.
43. Laidler, D. A.; Stoddart, J. F. *J. Chem. Soc., Chem. Commun.* **1977**, 481.
44. Laidler, D. A.; Stoddart, J. F.; Wolstenholme, J. B. *Tetrahedron Lett.* **1979**, 465.
45. Pettman, R. B.; Stoddart, J. F. *Tetrahedron Lett.* **1979**, 457, 461.
46. Laidler, D. A.; Stoddart, J. F. *Tetrahedron Lett.* **1979**, 453.
47. Bradshaw, J. S.; Baxter, S. L.; Scott, D. C.; Lamb, J. D.; Izatt, R. M.; Christensen, J. J. *Tetrahedron Lett.* **1979**, 3383.
48. Bradshaw, J. S.; Maas, G. E.; Izatt, R. M.; Lamb, J. D.; Christensen, J. J. *Tetrahedron Lett.* **1979**, 635.
49. Baxter, S. L.; Bradshaw, J. S. *J. Heterocyclic Chem.* **1981**, *18*, 233.
50. Colquhoun, H. M.; Stoddart, J. F. *J. Chem. Soc., Chem. Commun.* **1981**, 612.
51. Colquhoun, H. M.; Stoddart, J. F.; Williams, D. J. *J. Chem. Soc., Chem. Commun.* **1981**, 847, 849, 851.
52. Metcalfe, J. C.; Stoddart, J. F.; Jones, G.; Crawshaw, T. H.; Gavuzzo, E.; Williams, D. J. *J. Chem. Soc., Chem. Commun.* **1981**, 432.

53. Metcalfe, J. C.; Stoddart, J. F. *J. Am. Chem. Soc.* **1977**, *99*, 8317.
54. Krane, J.; Aune, O.; *Acta Chem. Scand.* **1980**, *34B*, 397.
55. Johnson, M. R. Ph.D. Thesis, Liverpool University, Liverpool, **1980**.
56. Bovill, M. J.; Chadwick, D. J.; Sutherland, I. O.; Watkin, D. *J. Chem. Soc., Perkin Trans. 2* **1980**, 1529.
57. Bovill, M. J.; Chadwick, D. J.; Johnson, M. R.; Jones, N. F.; Sutherland, I. O.; Newton, R. F. *J. Chem. Soc., Chem. Commun.* **1979**, 1065.
58. Jones, N. F. Ph.D. Thesis, Liverpool University, Liverpool, 1981.
59. Moore, S. S.; Tarnowski, T. L.; Newcomb, M.; Cram, D. J.; *J. Am. Chem. Soc.* **1977**, *99*, 6398.
60. Newcomb, M.; Moore, S. S.; Cram, D. J. *J. Am. Chem. Soc.* **1977**, *99*, 6405.
61. Lehn, J. M.; Vierling, P. *Tetrahedron Lett.* **1980**, 1323.
62. Cram, D. J.; Cram, J. M. *Acc. Chem. Res.* **1978**, *11*, 8.
63. Stoddart, J. F. *Chem. Soc. Rev.* **1979**, *8*, 85.
64. Lehn, J. M. *Struct. Bonding* (Berlin) **1973**, *16*, 1; *Acc. Chem. Res.*, **1978**, *11*, 49.
65. Cheney, J.; Kintzinger, J. P.; Lehn, J. M. *Nouv. J. Chim.* **1978**, *2*, 411.
66. Lehn, J. M.; Simon, J. *Helv. Chim. Acta* **1977**, *60*, 141.
67. Lehn, J. M.; Simon, J.; Moradpour, A. *Helv. Chim. Acta* **1978**, *61*, 2407.
68. Calverley, M. J.; Dale, J. *J. Chem. Soc., Chem. Commun.* **1981**, 1084.
69. Buhleier, E.; Wehner, W.; Vögtle, F. *Chem. Ber.* **1979**, *112*, 546, 559.
70. Jackman, L. M.; Sternhell, S. In "Applications of NMR Spectroscopy in Organic Chemistry"; Pergamon: New York, 1969; pp. 94–98.
71. Johnson, M. R.; Sutherland, I. O.; Newton, R. F. *J. Chem. Soc., Chem. Commun.* **1979**, 309.
72. Mageswaran, R.; Mageswaran, S.; Sutherland, I. O. *J. Chem. Soc., Chem. Commun.* **1979**, 722.
73. Jones, N. F.; Kumar, A.; Sutherland, I. O. *J. Chem. Soc., Chem. Commun.* **1981**, 990.
74. Kotzyba-Hibert, F.; Lehn, J. M.; Vierling, P. *Tetrahedron Lett.* **1980**, 941.
75. Kintzinger, J. P.; Kotzyba-Hibert, F.; Lehn, J. M.; Pagelot, A.; Saigo, K. *J. Chem. Soc., Chem. Commun.* **1981**, 833.
76. Behr, J. P.; Lehn, J. M. *J. Am. Chem. Soc.* **1976**, *98*, 1743.
77. Kotzyba-Hibert, F.; Lehn, J. M.; Saigo, K. *J. Am. Chem. Soc.* **1981**, *103*, 4266.
78. Fuller, S. E.; Mann, B. E.; Stoddart, J. F. *J. Chem. Soc., Chem. Commun.* **1982**, 1096.

2

NMR STUDIES IN LIQUID CRYSTALS: DETERMINATION OF SOLUTE MOLECULAR GEOMETRY

P. Diehl and J. Jokisaari*

DEPARTMENT OF PHYSICS
UNIVERSITY OF BASEL
SWITZERLAND

Introduction

The nuclear chemical shift and indirect spin–spin coupling are NMR parameters that furnish information on the electronic structures of molecules. Both depend upon the direction of molecular axes with respect to applied magnetic field; ie, they are anisotropic and must be described as tensors. In isotropic solutions, because of Brownian motion, these tensors are averaged over all molecular directions and only one third of the trace is observed as the resulting scalar parameter. Consequently, much useful information is lost.

Brownian motion is also responsible for the averaging of a third type of parameter, the direct spin–spin coupling, a magnetic dipole–dipole interaction that is, as opposed to indirect coupling, transmitted through space. For this interaction the average value in an isotropic solvent is zero, although the instantaneous value may correspond to several kilocycles of Larmor frequency and consequently contribute substantially to the nuclear relaxation.

Direct coupling has been observed and studied in solids since the early days of NMR spectroscopy in the 1950s. It has been used to derive information

*On leave from the Department of Physics, University of Oulu, Finland.

NMR in Stereochemical Analysis

about the structure of crystals. As it depends upon the inverse cube of the internuclear distance, it can measure, for example, proton–proton distances. The early method had one main disadvantage: because of the interactions of many neighboring nuclei, the linewidth in solid-state NMR was usually large. Consequently, the precision of the information obtained was not very high.

Therefore, about 20 years ago, NMR spectroscopists began to search for ways to orient molecules by various means so that the intermolecular dipole–dipole coupling could be averaged out to zero, as in isotropic liquids, but the coupling within the molecule would average to a nonzero value. It should then be feasible to observe the direct coupling in high-resolution spectroscopy.

At first, a variety of ideas was tested, for instance, orienting by electric fields,[1,2] absorption of molecules in polymers with subsequent stretching, or embedding molecules in zeolite crystal channels.[3] There was very restricted success; not until 1963, when Saupe and Englert[4,5] used liquid crystals as solvents for the first time, was the NMR of oriented molecules born, and it received immediate attention.

In liquid crystalline solutions the observed NMR parameters of solute molecules are averages for an anisotropic distribution of angles between molecular axes and liquid crystal axes, respectively, the applied magnetic field. This inhomogeneous distribution admits the study of the directional dependence of indirect spin–spin couplings or chemical shifts. More important, however, are the direct dipole–dipole interactions, which do not vanish as they do in isotropic cases. They can be described as products of the so-called degree of order or degree of orientation of the axis passing through the interacting nuclei (which characterizes the anisotropic distribution of molecular axes), and of the inverse third power of the internuclear distance. These direct couplings may be used to determine molecular structures and orientations.

In the early days of oriented molecule NMR, studies were confined mainly to the nuclei ^{1}H and ^{19}F, which exhibited large direct coupling constants. The spectral lines were generally broad and the precision of structure determination was small. Today, with Fourier transformation (FT) instruments and improved liquid crystal solvents, ^{1}H linewidths of 1 Hz may be achieved. Furthermore ^{13}C and ^{15}N satellites may be observed at their natural abundances in ^{1}H spectra and complete molecular structures can be studied.

The increased precision, on the other hand, also introduced new requirements and difficulties. An example is the molecular force fields for harmonic vibrations, which must be known in order to average the direct dipolar couplings over the intramolecular vibrational motion.

This chapter focuses on the use of NMR to determine molecular structure in partially oriented molecules, and on its benefits and limitations. Readers interested in more details of the resulting molecular structures, bond lengths, bond angles, and order parameters should consult the books[6,7] and many review articles[8–18] that have been published in this field.

Liquid Crystals

General Properties

Liquid crystals have been known for a long time. They were first observed in 1888.[19] A certain organic substance (cholesteryl benzoate) seemed to have two distinct melting points. At 145°C the solid turned into a cloudy liquid, which at 179°C became clear. The cloudy intermediate was soon shown to have a crystal-like structure and the name "liquid crystal" was suggested.

The complex structure and peculiar time dependence of the liquid crystal defects, as observed in polarized light, were misinterpreted as signs of life—perhaps of a precursor of life. Actually as modern research has shown, liquid crystals indeed play a very active role in life, although one very different from earlier suggestions, by forming the cell membranes.

Liquid crystal research reached a first peak in the 1930s and very famous physicists were attracted by this field (Bragg, de Broglie, Born). Later on, however, it seemed that all the major problems were solved and there appeared no practical possibility of application. The field lost attention and only in the 1960s did new interest arise.

Liquid crystals are composed of nonspherical, either rod- or disk-shaped molecules that often enclose aromatic rings, eg, azoxyanisole. The intermolecular forces tend to orient these molecules with their long axes parallel for the rods and the short axes parallel for the disks. In spite of this "local order", the liquid crystal molecules still rotate and diffuse quite similarly to isotropic liquids, and the liquid crystal has macroscopic properties, such as flow that is liquid-like.

There are three main types of liquid crystals: nematic, smectic, and cholesteric, which differ in their types of ordering. Nematic crystals have no long-range ordering in the centers of gravity of the molecules, but there is an order in the direction of molecular axes, which, if oriented homogeneously, for instance, by an applied magnetic field, constitute an optically uniaxial medium. The cholesterics are similar to nematics but their axes of local orientation form helical structures. The smectics, have a layer structure which is without long-range ordering of centers of gravity in the layer for types A and C, but with order for type B.

The most important for NMR spectroscopy are the nematic and the smectic A phases, which may be either mixtures or pure compounds that become nematic when the solid is melted (thermotropic), or alternatively mixtures of two or more compounds (lyotropic) of which one must be amphiphilic and the other usually is water. Lyotropic liquid crystals are formed because the amphiphilic has an ionic hydrophilic head that orients toward the water and a hydrophobic tail of a hydrocarbon chain, which avoids the water region.

The nematic phase of liquid crystals can be described theoretically, eg, by the Maier-Saupe mean field theory, which represents the simplest long-range order approach.[20] In analogy to Weiss' theory of ferromagnetism the intermolecular interactions responsible for the orientation of the individual molecules are characterized as an internal field; ie, each molecule has a potential energy $W=f(S,V,\Theta)$, which depends upon the momentary angle of the long molecular axis (Θ), the degree of order ($S=\frac{1}{2}\langle 3\cos^2\Theta-1\rangle$), and upon the molar volume (V). The anisotropy of the molecule perpendicular to the long axis is neglected; ie, there is only one degree of order characterizing the orientation. On this basis the result of Boltzmann statistics for the degree of order of the liquid crystal is:

$$S=\frac{3}{2}\frac{\int_0^{\pi/2}\exp(-W/kT)\cos^2\Theta\sin\Theta\, d\Theta}{\int_0^{\pi/2}\exp(-W/kT)\sin\Theta\, d\Theta}-\frac{1}{2} \qquad (2\text{-}1)$$

Consequently S as a function of temperature T and molar volume is defined if the potential W is known explicitly.

For W, Saupe and Maier chose a very simple relation based upon the assumption that dispersion interaction is predominantly responsible for the orientation. If only the pure dipole–dipole contribution to the dispersion interaction is considered, then:

$$W\approx -1/2\frac{S}{V^2}(3\cos^2\Theta-1) \qquad (2\text{-}2)$$

Although this simple theory cannot account for all the quantitative aspects of the nematic state, it qualitatively approximates various properties, such as the temperature variation of the degree of order.

Liquid crystals can easily be oriented in magnetic fields of the order of 0.1 T. The molecular aggregates will, as a consequence of their anisotropic diamagnetism, orient their axes of largest susceptibility perpendicular to the field. For liquid crystal molecules enclosing aromatic rings, this axis is perpendicular to the rings. Consequently these molecules orient with their long axes along the applied magnetic field. The resulting optic axis of the liquid crystal is perpendicular to the sample axis if a nonsuperconducting magnet is used, and sample spinning to obtain higher resolution of the NMR spectrum is not possible. For liquid crystals with axes of high diamagnetism along the long molecular axis the situation is, of course, opposite; spinning of samples is possible in conventional magnets but not in superconducting ones.

In liquid crystals of the smectic A type, the optic axis orients normally along the magnetic field, if the sample is cooled from the isotropic state in the magnetic field. The molecular order being quite stable, the sample can subsequently be rotated to any angle between the optic axis and the applied field and kept stable for several hours.

Solute Orientation

Molecules that are dissolved in oriented liquid crystals also become oriented by interaction with the solvent. The degree of order of the solute is a function of the strength of this interaction as well as of the temperature and solute concentration. A rough estimate of variation is as follows:

The relative change of orientation is of the order of −15 to −30% for an increase of solute concentration by a factor of two.

The relative change of orientation is of the order of −1% per degree increase in sample temperature.

In analogy with the Maier-Saupe theory, the degrees of order of the solute molecules in liquid crystal solvents may be characterized by the anisotropic mean field potential originating in the dispersion interaction experienced by a solute molecule surrounded by solvent molecules. With the simplifying assumption that the solvent molecules are cylindrically symmetrical and neglecting the orientation-independent terms, one obtains for the potential:[21]

$$W = -C\rho\Delta\alpha\ S\,(\alpha_{YY} - \alpha_{XX})\,(\sin^2\Theta\,\sin^2\phi + \mathrm{R}\cos^2\Theta) \qquad (2\text{-}3)$$

where $\Delta\alpha = \alpha_{\parallel} - \alpha_{\perp}$ is the anisotropy of the liquid crystal molecular polarizability and the $\alpha_{\mu\mu}$ are the principal polarizabilities of the solute molecules.

$$C = \frac{1}{20}\cdot\frac{1}{(4\pi\varepsilon_0)^2}\cdot\frac{\varepsilon_1 . \varepsilon_2}{\varepsilon_1 + \varepsilon_2}\cdot 4\pi\int_0^\infty \frac{g(r_{12})}{r_{12}^4}\,dr_{12} \qquad (2\text{-}4)$$

where ε_ν is the energy parameter of the molecule ν; $g(r_{12})$ is the radial distribution function of the liquid crystal molecules around the solute molecule; Θ and ϕ are the polar angles of the optic axis in the molecular coordinate system and:

$$R = (\alpha_{zz} - \alpha_{xx})/(\alpha_{yy} - \alpha_{xx}) \qquad (2\text{-}5)$$

On this basis the ordering matrix elements S_{xx} and S_{zz} with respect to the optic axis are derived as follows:

$$S_{xx} = \frac{3}{2Z}\int \sin^2\Theta\,\cos^2\phi\,\exp\,(-W/kT)\,d\Omega - \frac{1}{2} \qquad (2\text{-}6)$$

$$S_{zz} = \frac{3}{2Z}\int \cos^2\Theta\,\exp\,(-W/kT)\,d\Omega - \frac{1}{2} \qquad (2\text{-}7)$$

with

$$Z = \int \exp\,(-W/kt)\,d\Omega \qquad (2\text{-}8)$$

A corresponding approach based upon repulsive intermolecular forces leads to similar expressions, but with all the $\alpha_{\mu\mu}$ replaced by R_μ, the molecular dimensions in the μ direction.

Solute order, of course, also may be determined by specific interactions with the solvent, particularly by hydrogen bonding.

Basic Theory

The Order Tensor

For the definition of the degree of order of a rigid molecule in a liquid crystal, Saupe[22] introduced the so-called order tensor, $\bar{\bar{S}}$:

$$S_{pq} = \frac{1}{2}\langle 3 \cos \Theta_{pz'} \cos \Theta_{qz'} - \delta_{pq} \rangle \tag{2-9}$$

where $\Theta_{pz'}$ is the angle between the liquid crystal optic axis and the p axis in a molecule's fixed coordinate system, δ_{pq} is the Kronecker delta, and the z′ axis coincides with the optic axis. This tensor is traceless and symmetrical. Therefore, it consists of five independent elements. This number may be reduced in many cases, bacause the $\bar{\bar{S}}$ tensor must be invariant for all the symmetry operations of a molecule. For example, if a molecule possesses a plane of symmetry, one of the principal axes of the tensor is perpendicular to this plane. If the c axis of the molecule is selected parallel to this axis, the order tensor elements S_{ac} and S_{bc} vanish. For a molecule with two perpendicular planes of symmetry, which contain the a and b axes, the three elements, S_{ab}, S_{ac}, and S_{bc}, are zero if the c axis is selected as the line of intersection of the two planes. There will be only one independent S element, S_{cc}, if a molecule has an n-fold ($n \geqslant 3$) symmetry axis. Then the order tensor is diagonal and the elements are related through the equation:

$$S_{aa} = S_{bb} = -\tfrac{1}{2} S_{cc} \tag{2-10}$$

It should be pointed out that the values of the order tensor elements, S_{pq}, usually reported in the literature are actually equal to $P_2(\cos \beta) S_{pq}$, where $P_2(\cos \beta) = \frac{1}{2}(3 \cos^2 \beta - 1)$ is the Legendre polynomial and β is the angle between the applied magnetic field and the liquid crystal optic axis.

Snyder[23] has introduced an alternative description of the molecular orientation in terms of motional constants, but in this chapter the $\bar{\bar{S}}$ tensor is used exclusively.

The degree of order of a molecular axis connecting the nuclei i and j can be expressed in terms of the order tensor as follows:

$$S_{ij} = \sum_{p,q} \cos \Theta_{ijp} \cos \Theta_{ijp} S_{pq} \tag{2-11}$$

where Θ_{ijp} is the angle between the ij axis and the p axis of the molecule's fixed coordinate system.

If the five independent elements of the order tensor are selected as S_{cc}, $(S_{aa} - S_{bb})$, S_{ab}, S_{ac}, and S_{bc}, S_{ij} can be written in the form:

$$\begin{aligned} S_{ij} = {} & \tfrac{1}{2} S_{cc} (3 \cos^2 \Theta_{ijc} - 1) + \tfrac{1}{2} (S_{aa} - S_{bb}) \\ & \cdot (\cos^2 \Theta_{ija} - \cos^2 \Theta_{ijb}) + 2 S_{ab} \cos \Theta_{ija} \cos \Theta_{ijb} \\ & + 2 S_{ac} \cos \Theta_{ija} \cos \Theta_{ijc} + 2 S_{bc} \cos \Theta_{ijb} \cos \Theta_{ijc} \end{aligned} \tag{2-12}$$

In every oriented molecule there is a continuum of directions for which the degree of order vanishes, i.e., the right-hand side of eq. (2-11) is zero:

$$a^2 S_{aa} + b^2 S_{bb} + c^2 S_{cc} + 2\,ab S_{ab} + 2\,ac S_{ac} + 2\,bc S_{bc} = 0 \qquad (2\text{-}13)$$

where the relation $\cos \Theta_{ijp} = p/r_{ij}$ ($p = a,\ b,\ c$) was used. Equation (2-13) describes a double cone and transforms into a relatively simple form, if a molecule with two planes of symmetry is regarded:

$$a^2 S_{aa} + b^2 S_{bb} + c^2 S_{cc} = 0 \qquad (2\text{-}14)$$

If the molecule lies in the *ab* plane ($c=0$), we find $a/b = \pm\sqrt{-S_{bb}/S_{aa}}$, or in the *ac* plane ($b=0$), $a/c = \pm\sqrt{-S_{cc}/S_{aa}}$. There is a direction therefore, for which $S=0$ in the molecular plane if the two S values have opposite signs. The problems originating in the zero or very small S values in molecular structure determinations are discussed in the sections on molecular vibration and correlation.

The degree of order of orientation of an axis α, S_α, is not identical to the probability $P(\Theta,\phi)$ that the direction of the α axis is (Θ,ϕ) in molecular spherical coordinates. However, $P(\Theta,\phi)$ can be expanded in a series of spherical harmonics, $Y_{l,m}(\Theta,\phi)$

$$P(\Theta,\ \phi) = \sum_{l,m} a_{l,m} Y_{l,m}(\Theta,\ \phi) \qquad (2\text{-}15)$$

and can be compared with the order tensor elements.

If the mesophase is not affected by strong electric fields, the orientation of the molecules is apolar. Then $P(\Theta,\phi)$ is invariant with respect to the inversion $(\Theta,\phi)\rightarrow(\pi-\Theta,\pi+\phi)$ and the series (2-15) only includes spherical harmonics with even l.

$$\begin{aligned}P(\Theta,\phi) = \tfrac{1}{4\pi}\,\{1 + 5/2[(3\cos^2\Theta - 1)S_{cc} + \sin^2\Theta\cos 2\phi\,(S_{aa} - S_{bb})\\ + 4\sin\Theta\cos\Theta\cos\phi\,S_{ac} + 4\sin\Theta\cos\Theta\sin\phi\,S_{bc}\\ + 2\sin^2\Theta\sin 2\phi\,S_{ab}]\} + \text{higher order terms}\end{aligned} \qquad (2\text{-}16)$$

Inspection of eq. (2-16) makes clear that, as measured by NMR, the elements of the order tensor of the oriented molecules characterize the angular distribution probability of molecular axes only up to terms with $l=0$ and $l=2$.

A comparison with the theoretical probability distribution $P(\Theta) = A\exp(-B\cos^2\Theta)$ (where B = interaction energy/kT) shows that the higher order terms, ie $l=4$, 6, ..., are important, particularly the term $l=4$. Consequently, the degrees of orientation are not a good measure of the angular distribution probability.

Direct Coupling

The direct dipolar coupling tensor $\bar{\bar{D}}'$ (in SI units and neglecting intramolecular motions) is defined as

$$D'_{ij\alpha\beta} = -\frac{\mu_0 \hbar \gamma_i \gamma_j}{8\pi^2 r_{ij}^3}(3\cos\Theta_{ij\alpha}\cos\Theta_{ij\beta} - \delta_{\alpha\beta}) \qquad (2\text{-}17)$$

where $\Theta_{ij\alpha}$ is the angle between the α axis and the vector $\bar{r}_{ij}$ from nucleus i to nucleus j, γ_i is the gyromagnetic ratio of the nucleus i, $\hbar$ is Planck's constant divided by 2π, and μ_0 the permeability of vacuum. The tensor $\bar{\bar{D}}'$ is traceless, and therefore $D'^{iso} = Tr\ \bar{\bar{D}}' = 0$, whereas $\langle D'_{ijzz} \rangle = D'^{aniso}_{ij}$. The dipolar coupling constant is usually defined as $D_{ij} = \frac{1}{2} D'^{aniso}_{ij}$. Consequently,

$$D_{ij} = -\frac{\mu_0 \hbar \gamma_i \gamma_j}{8\pi^2} \sum_{\alpha,\beta} \left\langle \frac{r_\alpha r_\beta}{r^5} \left(\frac{3}{2} \cos \Theta_{\alpha z} \cos \Theta_{\beta z} - \frac{1}{2} \delta_{\alpha\beta} \right) \right\rangle \tag{2-18}$$

where the average is taken over the inter- and intramolecular motion. If the molecular structure and orientation can be assumed independent (no correlation; see the section on correlation), then eq. (2-18) can be rewritten:

$$D_{ij} = -\frac{\mu_0 \hbar \gamma_i \gamma_j}{8\pi^2} \sum_{\alpha\beta} \left\langle \frac{r_\alpha r_\beta}{r^5} \right\rangle S_{\alpha\beta} \tag{2-19}$$

The degree of orientation of the ij axis is

$$S_{ij} = \tfrac{1}{2} \langle 3 \cos^2 \Theta_{ijz} - 1 \rangle \tag{2-20}$$

Equation (2-18) together with eq. (2-12) constitute the basis for the determination of molecular structures.

The Hamiltonian

To analyze the NMR spectrum of a molecule dissolved in a liquid crystal, the form of the Hamiltonian operator must be known. We can write

$$\hat{H} = \hat{H}_Z + \hat{H}_J + \hat{H}_D + \hat{H}_Q \tag{2-21}$$

where $\hat{H}_Z$ is the operator for the Zeeman interaction (interaction of a nuclear magnetic dipole with the magnetic field), $\hat{H}_J$ takes into account the indirect spin–spin interactions (transfered through electrons), $\hat{H}_D$ describes the direct dipole–dipole interactions (transfered through space), and $\hat{H}_Q$ is required for nuclei with spin $I > \frac{1}{2}$ and represents the quadrupole interaction (interaction of the nuclear quadrupole moment with the electric field gradient).

The electron system of a molecule in a magnetic field induces local fields at the nuclei. The effective field experienced by a nucleus differs from the external field, in both size and direction, and consequently this effect (shielding) must be described as a second-rank tensor, $\bar{\bar{\sigma}}$, the shielding tensor. With $\bar{\bar{\sigma}}$ the operator $\hat{H}_Z$ can be written (in Hz) as

$$\hat{H}_Z = -\frac{1}{2\pi} \sum_i \gamma_i \hat{\mathbf{I}}_i \cdot (\bar{\bar{\mathbf{1}}} - \bar{\bar{\sigma}}_i) \cdot \bar{\mathbf{B}} \tag{2-22}$$

where $\bar{\mathbf{B}}$ is the external field and $\bar{\bar{\mathbf{1}}}$ is the unit matrix. According to common practice the z axis of the laboratory frame is chosen along the field direction. A good approximation for $\hat{H}_Z$ can be obtained by considering only the term:

$$\hat{H}_Z = -\frac{1}{2\pi}\sum_i \gamma_i \hat{I}_{iz}(1-\sigma_{izz})B_z \qquad (2\text{-}23)$$

which commutes with $\sum_i \hat{I}_{iz}$, σ_{izz} is the z component of the shielding tensor in the laboratory frame. In an NMR experiment with oriented molecules, the mean value $\langle\sigma_{izz}\rangle$ is obtained, which can be presented in the form:

$$\langle\sigma_{izz}\rangle = \sigma_i^{iso} + \sigma_i^{aniso} \qquad (2\text{-}24)$$

Here σ_i^{iso} is the shielding constant observed in an isotropic liquid, whereas $\sigma_i{}^{aniso}$ is the anisotropic contribution.

The indirect spin–spin Hamiltonian can be written in general form:

$$\hat{H}_J = \sum_{i<j} \hat{I}_i \cdot \bar{\bar{\mathbf{J}}}_{ij} \cdot \hat{I}_j \qquad (2\text{-}25)$$

where $\bar{\bar{\mathrm{J}}}_{ij}$ is the second-rank coupling tensor. As in the case of the Zeeman interaction, here also the terms that do not commute with the total spin component I_z can be neglected. Then the approximate Hamiltonian is

$$\hat{H}_J = \sum_{i<j}\left[J_{ijzz}\hat{I}_{iz}\hat{I}_{jz} + \frac{1}{4}(J_{ijxx}+J_{ijyy})(\hat{I}_{i+}\hat{I}_{j-}+\hat{I}_{i-}\hat{I}_{j+}) + \frac{i}{4}(J_{ijxy}-J_{ijyx})(\hat{I}_{i+}\hat{I}_{j-}-\hat{I}_{i-}\hat{I}_{j+})\right] \qquad (2\text{-}26)$$

where $\hat{I}_+ = \hat{I}_x + i\hat{I}_y$ and $\hat{I}_- = \hat{I}_x - i\hat{I}_y$ are the raising and lowering spin operators, respectively. Because only average values of the components of $\bar{\bar{\mathrm{J}}}_{ij}$ are observed, the last term in eq. (2-26) vanishes. Furthermore,

$$\tfrac{1}{4}(\langle J_{ijxx}\rangle + \langle \mathrm{J}_{ijyy}\rangle) = \tfrac{1}{2}J_{ij}^{iso} - \tfrac{1}{4}J_{ij}^{aniso} \qquad (2\text{-}27)$$

where J_{ij}^{iso} is the conventional spin–spin coupling constant measured in isotropic liquids. Substituting $\langle J_{ijzz}\rangle = J_{ij}^{iso} + J_{ij}^{aniso}$ in eq. (2-26), gives:

$$\hat{H}_J = \sum_{i<j}\left\{J_{ij}^{iso}\left[\hat{I}_{iz}\hat{I}_{jz} + \frac{1}{2}(\hat{I}_{i+}\hat{I}_{j-}+\hat{I}_{i-}\hat{I}_{j+})\right] + J_{ij}^{aniso}\left[\hat{I}_{iz}\hat{I}_{jz} - \frac{1}{4}(\hat{I}_{i+}\hat{I}_{j-}+\hat{I}_{i-}\hat{I}_{j+})\right]\right\} \qquad (2\text{-}28)$$

The term $\hat{H}_D$ in eq. (2-21) describes the direct interaction between nuclear dipole moments. Classically it is written (in Hz):

$$\hat{H}_D = -\frac{\mu_0\hbar}{8\pi^2}\sum_{i<j}\gamma_i\gamma_j\left[\frac{3(\mathbf{I}_i\cdot\bar{\mathbf{r}}_{ij})(\mathbf{I}_j\cdot\bar{\mathbf{r}}_{ij})}{\mathrm{r}_{ij}^5} - \frac{\mathbf{I}_i\cdot\mathbf{I}_j}{r_{ij}^3}\right] \qquad (2\text{-}29)$$

where $\bar{\mathrm{r}}_{ij}$ is the vector from the nucleus i to nucleus j. In fact, the definition (2-29) is identical to eq. (2-25) if $\bar{\bar{\mathrm{J}}}_{ij}$ is replaced by the direct dipole–dipole coupling tensor $\bar{\bar{\mathrm{D}}}'_{ij}$. Consequently, using the definition $D_{ij} = \frac{1}{2}\, D'^{aniso}_{ij}$ (see the section, Direct Coupling, eq. (2-30) can be derived:

$$\hat{H}_D = \sum_{i<j} 2D_{ij}\left[\hat{I}_{iz}\hat{I}_{jz} - \frac{1}{4}(\hat{I}_{i+}\hat{I}_{j-} + \hat{I}_{i-}\hat{I}_{j+})\right] \tag{2-30}$$

The direct and indirect coupling Hamiltonians together give:

$$\hat{H}_J + \hat{H}_D = \sum_{i<j} (J_{ij} + 2D_{ij}^{exp})\hat{I}_{iz}\hat{I}_{jz} + \frac{1}{2}\sum_{i<j} (J_{ij} - D_{ij}^{exp})(\hat{I}_{i+}\hat{I}_{j-} + \hat{I}_{i-}\hat{I}_{j+}) \tag{2-31}$$

where $D_{ij}^{exp} = D_{ij}^{dir} + D_{ij}^{indir} = D_{ij}^{dir} + \frac{1}{2}J_{ij}^{aniso}$. Using the general coupling tensor $\bar{\bar{T}}_{ij}$ ($T_{ij}^{iso} = J_{ij}^{iso}$, $T_{ij}^{aniso} = J_{ij}^{aniso} + 2D_{ij}^{dir}$) we find:

$$\hat{H}_T = \hat{H}_J + \hat{H}_D = \sum_{i<j}\left\{T_{ij}^{iso}\left[\hat{I}_{iz}\hat{I}_{jz} + \frac{1}{2}(\hat{I}_{i+}\hat{I}_{j-} + \hat{I}_{i-}\hat{I}_{j+})\right] + T_{ij}^{aniso}\left[\hat{I}_{iz}\hat{I}_{jz} - \frac{1}{4}(\hat{I}_{i+}\hat{I}_{j-} + \hat{I}_{i-}\hat{I}_{j+})\right]\right\} \tag{2-32}$$

From eq. (2-31), it can be seen that the "anisotropic" Hamiltonian operator is obtained from the corresponding "isotropic" operator simply by replacing J_{ij} by $(J_{ij} + 2D_{ij}^{exp})$ in the terms producing diagonal elements of the Hamilton matrix, and by $(J_{ij} - D_{ij}^{exp})$ in the off-diagonal terms. Therefore, the analysis of the spectra of partially oriented molecules is quite similar to that of isotropic systems. It should be pointed out however, that the equivalence of spins no longer has the same meaning as in isotropic systems; ie, the couplings between equivalent nuclei affect a spectrum. For example, the six protons of benzene give a very rich spectrum.

If the resonating nucleus possesses a spin $I \geqslant 1$, the term $\hat{H}_Q$ must be included in the total Hamiltonian:

$$\hat{H}_Q = \frac{1}{h}\sum_i \frac{eQ_i\langle V_{izz}\rangle}{4I_i(2I_i - 1)}[3\hat{I}_{iz}^2 - I_i(I_i + 1)] \tag{2-33}$$

Here eQ_i is the nuclear quadrupole moment and $\langle V_{izz}\rangle$ the average value of the electric field gradient; the spins are considered quantized along the magnetic field direction and the quadrupole coupling is assumed weak compared to the magnetic interaction.

Experimental Use

Instrumentation

The first NMR spectra of partially oriented molecules were recorded on continuous-wave (CW) spectrometers. The resulting linewidths typically ranged from 4 to 60 Hz, with relatively sharp lines in the center of the spectrum and broad lines in the wings. As pointed out above, with modern Fourier transform (FT) spectrometers, good sample-temperature-controlling instruments, and improved liquid crystals that are nematic at room temperature, linewidths of the order of 1 Hz throughout the whole spectral range may be accomplished. The sensitivity furthermore permits the observation of ^{13}C and ^{15}N satellites at natural abundance in ^{1}H spectra. Depending upon the orientation of the liquid crystal axis (see the section, Liquid Crystals Used) samples can be rotated either in the conventional electromagnets or in superconductive magnets. This also reduces the linewidths.

Satellite observation significantly amplifies the information for molecular structure determination. However, the low gyromagnetic ratios of, eg, ^{13}C and ^{15}N responsible for the small dipolar coupling constants sometimes cause difficulties. The satellites may appear as shoulders on the much stronger normal proton lines (arising from molecules containing ^{12}C or ^{14}N) or may be completely hidden. For the suppression of the strong proton transitions, a method called abundant isotope elimination by fourier transform (AISEFT) was proposed.[24] The basic idea of the method is to record two free induction decay (FID) signals, the first one under normal conditions and the second with decoupling irradiation at the frequency of the rare nucleus (^{13}C or ^{15}N or any other). Subsequently, the Fourier transformation of the difference of the two FIDs is calculated. The resulting spectra, as shown in Figure 2-1, contain the satellites caused by the rare nuclei exclusively as positive signals, as well as a negative signal at the position of the disappeared strong line with intensity equal to the sum of the satellites. The process can, of course, be repeated as many times as required for the achievement of a reasonable signal to noise ratio.

On FT spectrometers it is also possible to detect directly the rare spins in oriented molecules. For example proton-coupled ^{13}C, ^{15}N, ^{111}Cd, ^{113}Cd, ^{117}Sn, ^{119}Sn, and ^{199}Hg NMR spectra of some small solute molecules in thermotropic nematic solvents have been observed.[25-30] For these nuclei the temperature homogeneity and stability of the sample is particularly important because the large chemical shift anisotropy (7500 ppm for ^{199}Hg in dimethyl mercury[29]) induces considerable line broadening for minute temperature gradients.

Proton decoupling sometimes may be useful in simplifying the ^{2}H or ^{13}C spectra of complex molecules. On the other hand, in many cases it is necessary to know the direct couplings to protons to make a correct determination of molecular order or of the structure. Direct coupling constants often enclose indirect contributions that are difficult to isolate.[31] This is true for the nuclear pairs ^{13}C-^{13}C, ^{13}C-^{19}F, ^{13}C-^{15}N, ^{13}C-^{31}P and ^{19}F-^{19}F.

In decoupling experiments it should be noted that a high decoupling power may increase the sample temperature, which in turn affects the molecular orientation, ie, the observed direct coupling constants.

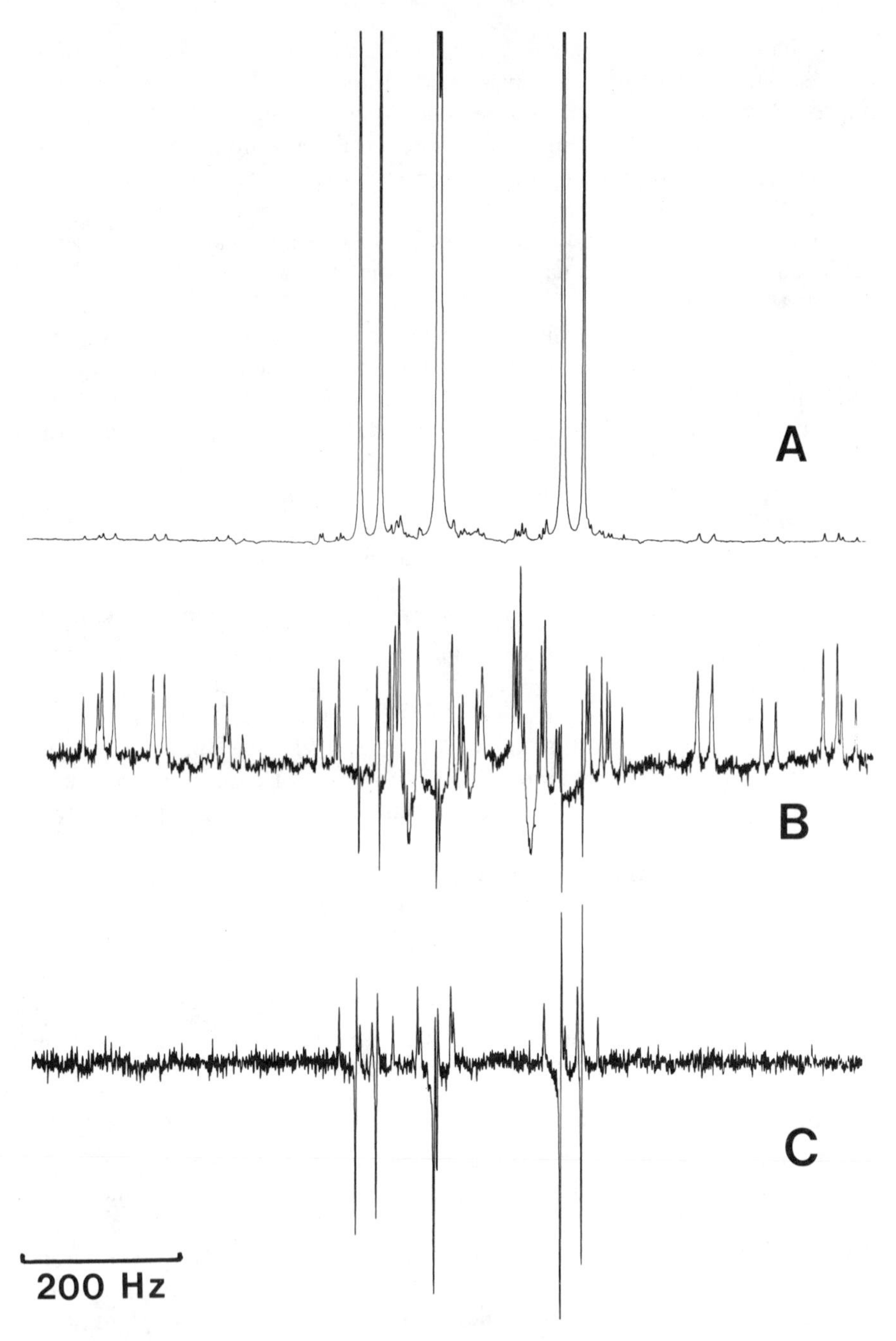

Figure 2-1. Abundant isotope elimination by Fourier transform (AISEFT): (A) ^{1}H spectrum of partially oriented *p*-chlorobenzonitrile (part of symmetrical spectrum); (B) Difference spectrum: $^{1}H-{}^{1}H\{{}^{13}C\} = {}^{13}C$ satellites; (C) Difference spectrum: $^{1}H-{}^{1}H\{{}^{15}N\} = {}^{15}N$ satellites.

Samples

Thermotropic liquid crystals allow a maximum solute concentration of about 20 mol %, whereas in lyotropic phases the usual concentration is approximately 1 wt %. Preferably 5-mm coaxial double-wall tubes with an inner diameter of approximately 2.7 mm and the locking substance in the interwall space are used for recording ^1H spectra. For lower sensitivity nuclei 10-mm sample tubes with an external lock are recommended for a maximum sample volume. In lyotropic solvents the internal D_2O may be used as a lock substance.

To obtain an optimum linewidth in the spectrum, the samples should be degased and then homogenized by heating to the isotropic state and subsequent shaking. This process is repeated several times, because the spectra of the oriented molecules are very sensitive to concentration gradients.

For the determination of molecular structure or spin–spin coupling anisotropy no reference compounds are needed. In studies of chemical shift anisotropies, in contrast, either internal or external reference substance are required. For ^1H and ^{13}C spectra ^{13}C-enriched methane can be added to the sample through a vacuum line.

This chapter does not discuss the measurement of shift anisotropy any further; the reader may refer to a recent review article.[31]

Liquid Crystals Used

A large number of organic substances exhibit liquid crystalline properties.[32] In the NMR spectroscopy of solute molecules (determinations either of structures or of nuclear chemical shift and spin–spin coupling anisotropies) the most commonly used thermotropic nematic liquid crystals are formed of benzylidene (MBBA n-(p-methoxybenzylidene)-p-butylaniline, EBBA n-(p-ethoxybenzylidene)-p-butylaniline), azobenzene, azoxybenzene (HOAB p,p′-di-n-hexyloxyazoxybenzene, Phase IV, Phase V, HAB p,p′-di-n-heptylazoxybenzene), phenylbenzoate, or benzoic acid derivatives. All these have positive anisotropies of diamagnetic susceptibility, $\Delta\chi$, and therefore orient so that the director (optic axis) is parallel with the applied magnetic field. Consequently, in a conventional electromagnet the director is perpendicular to the axis of the sample tube and prevents sample rotation. In contrast, in a superconducting magnet the sample tube can be rotated. In 1980, a new liquid crystal, Merck ZLI 1167, which is a mixture of 4-*N*-alkyl-*trans,trans*-bicyclohexyl-4′-carbonitriles and possesses a negative diamagnetic anisotropy (see Table 2-1), was introduced. This particular liquid crystal allows sample rotation in a conventional magnet and makes it possible to reach ^1H linewidths of even better than 1 Hz.

Besides thermotropic nematic liquid crystals, a great variety of lyotropic nematic liquid crystals, formed by a mixture of C_8 or C_{10} alkyl sulfates and the corresponding alcohols, sodium sulfate, and water, have been used as orienting solvents.[12] Such solutions in general are fairly viscous and must be kept in the magnetic field for a long time before good linewidths can be achieved.

TABLE 2-1. PHYSICAL PROPERTIES OF SOME THERMOTROPIC NEMATOGENS COMMONLY USED IN NMR SPECTROSCOPY

Code name (manufacturer)	Composition and structure	Nematic range (°C)	Viscosity ($mm^2 s^{-1}$)	Dielectric permittivity	Dielectric anisotropy	Diamagnetic anisotropy
EBBA (Eastman)	*n*-(*p*-Ethoxybenzylidene)-*p*-butylaniline, C_2H_5O–⟨O⟩–$CH=N$–⟨O⟩–C_4H_9	35–79		4.8 (25°C)	−0.32 (25°C)	15.9 (22°C)
Phase IV (Merck)	Eutectic mixture of *p*-methoxy-*p*′-*n*-butylazoxybenzenes, CH_3O–⟨O⟩–$N=N$–⟨O⟩–CH_4H_9	20–74	34 (20°C)	5.6 (20°C)	−0.2 (20° C)	15.0 (22°C)
S1114 (Merck)	*trans*-4-Pentyl-(4-cyano-phenyl) cyclohexane, C_5H_{11}–⟨H⟩–⟨O⟩–CN	30–55	21 (20°C)	8.4 (20°C)	9.9 (20°C)	4.0 (22°C)
ZLI 1167 (Merck)	Mixture of 4-*n*-alkyl-*trans*,*trans*-bicyclohexyl-4′-carbonitriles, 36% C_3H_7–⟨H⟩–⟨H⟩–CN 34% C_5H_{11}–⟨H⟩–⟨H⟩–CN 30% C_7H_{15}–⟨H⟩–⟨H⟩–CN	32–83	30 (20°C)	6.1 (20°C)	3.9 (35°C)	−4.5 (22°)

Furthermore, the concentration of the solute molecules usually must be small (approximately 1 wt %). Consequently, the information of the weak ^{1}H–^{n}X satellite spectra may be lost. In addition, the solute molecules in lyotropic phases are not as well oriented as in thermotropic phases.

Table 2-1 gives the compositions and physical properties of a few thermotropic liquid crystals commonly applied in NMR spectroscopy. Because of the possible specific solute–solvent interactions (complex formation, hydrogen bonding, etc; see section on Multiple Sites), not all the liquid crystals are "equally good solvents". Therefore, the use of several solvents of various types is recommended in order to make sure that undistorted molecular structures are determined.

Spectral Analysis

Computer Programs

For oriented systems of nuclei with spin $I=\frac{1}{2}$ the analysis of the NMR spectra is based on the use of the Hamiltonian's given in eqs. (2-23) and (2-32). For spins $I\geqslant 1$ a term that describes the quadrupole coupling energy must be added.

In principle the analysis of the spectra of solute molecules oriented in the liquid crystal solvent is similar to that of the spectra of isotropic liquids. There are, however, a few differences.

First, as pointed out above, there are more types of parameters (chemical shifts, indirect couplings, direct couplings, and quadrupole couplings). As a consequence, only the line positions of an *AB* spectrum, for example, do not define fully the three parameters σ_{AB}, J_{AB}, and D_{AB}, also the intensities are required.

Second, in contrast to the indirect couplings, the direct couplings are also observable between magnetically equivalent nuclei. For example, an A_2 spin system gives a doublet spectrum, the peaks being separated by $3|D_{AA}|$. In general n equivalent spin-$\frac{1}{2}$ nuclei produce a spectrum of n lines equally separated by $3|D_{AA}|$ and with intensity ratios given by the binomial coefficients. For more details see the existing literature.[6,7]

Third, because of the large direct dipole–dipole coupling constants ($\approx 10^3$ Hz) the limit of weak interaction (*X* approximation) is normally reached only if the interacting nuclei are of different types, for example, ^{1}H and ^{13}C.

The spectra of oriented molecules are analyzed by use of computers. There exist several programs for this purpose, such as UEA,[33] and LEQUOR,[34] modified from LAOCOON[35] to include dipole–dipole and quadrupolar interactions. A corresponding version, PANIC,[36] is also available and can be used in a time-sharing mode on an ASPECT 2000 computer of a commercial FT NMR spectrometer.

The usual way to begin the analysis of a spectrum is to look for a set of good starting parameters (ν_i, J_{ij}, D_{ij}, B_{ii}) that produce a theoretical spectrum allowing the assignment of at least a few lines.[37] In the progress of the analysis the number of assigned lines is increased in steps to the maximum. The first-guess D value can be obtained from an approximately known molecular structure and estimated order parameters. The indirect coupling constants are taken from the "isotropic" spectrum and are usually kept constant in the iteration process.

If the molecular structure is not well known and the number of order parameters is large, the first estimation of starting values may become very difficult or impossible. For such situations, automatic programs for spectral analysis have been developed. The two existing versions, STREAK[38,39] and DANSOM,[40] are based on similar principles and can find the solution from practically arbitrary starting values.

The analysis of complex proton systems may be simplified by partial deuteration and decoupling of the deuterons during the recording of the ^{1}H spectra.[41] For these experiments it must be kept in mind that the substitution of deuterons for protons may affect the S tensor elements.

The available computer programs normally include the error calculation for the iterated parameters.[37] The errors of the parameters $\bar{p}$ are completely defined by the variance covariance matrix $\bar{\bar{C}}_p$, belonging to the parameter set $\bar{p}$. The square roots of the diagonal elements of $\bar{\bar{C}}_p$ are the standard deviations σ_i of the p_i, whereas the off-diagonal elements are measures of the interdependence of errors (or parameters). For the determination of errors, an average line-position error δ must be fed into the program. This error is used if the root-mean-square (rms) error of the fit is smaller than δ.

The error of the line positions in turn depends upon the linewidth $\nu_{\frac{1}{2}}$, the signal to noise ratio, S/N, and the point resolution, ν_p, of a spectrum as follows:

$$\delta = [\nu_{1/2}^2/(S/N) + \nu_p^2/12]^{1/2} \qquad (2\text{-}34)$$

In general the direct coupling constants D are interdependent. For large covariances it is recommended that combinations of Ds be used in the iterative process that determines the molecular structure (see the section, Computer Program for Structural Determination).[42] These combinations are obtained by diagonalization of the variance–covariance matrix:

$$p'_j = \sum_k^n a_{kj} p_k \qquad (2\text{-}35)$$

where n is the number of iterated parameters. The linear combinations p'_j either maximize or minimize the corresponding errors.

The Use of Satellites

A normal ^{1}H spectrum of an oriented molecule provides information on the proton skeleton (and order tensor elements) exclusively. The simultaneous observation of the weak ^{1}H–^{13}C satellites also defines the carbon skeleton.

Furthermore, small proton systems that do not carry any structural information may be analyzed in terms of molecular structure if the satellites are included.[43-45] As pointed out in the section on instrumentation the low natural abundance and small gyromagnetic ratios may cause problems in the detection of satellites. In order to avoid these limitations, the AISEFT method was developed (see section on Instrumentation).

In a few simple cases, ^1H–^nX satellite spectra can be analyzed "by hand" because certain splittings yield the quantities $|J_{HX} + 2D_{HX}|$. In general, however, satellite spectra also must be analyzed with computers as described in the section on Spectral Analysis, Computer Programs.

Satellite spectra often are quite complex because they are superpositions of spectra stemming from molecules with one nucleus, X, exclusively in each possible X position. With several equivalent X positions the increased satellite intensity may help to identify the corresponding isotopomer. Obviously the samples should be completely pure in order to avoid confusing small impurity peaks.

As mentioned in the previous chapter, the precision of a direct coupling constant is related to the signal-to-noise ratio of the transitions. In ^1H–^nX spectra, this ratio is often two orders of magnitude smaller than in pure ^1H spectra. As a consequence, the errors in D_{HX} are roughly five to ten times larger than those of the D_{HH} couplings.

On the other hand, the signal-to-noise ratio of a satellite is always larger than that of the corresponding transition in the X signal. In other words, as long as only couplings D_{HX} are of interest, it is always preferable to record the satellite in the ^1H spectrum. The gain in signal to noise ratio that may be achieved is summarized in Table 2-2.

TABLE 2-2. "DIRECT" AND "INDIRECT" RELATIVE RECEPTIVITIES OF SOME SPIN-$\frac{1}{2}$ NUCLEI

Nucleus	Natural abundance (%)	"Direct" receptivity,[a] R_X^H	"Indirect" receptivity[b] I_X^H	Gain factor[c]
^1H	99.985	1.000	1.000	
^{13}C	1.108	1.76×10^{-4}	1.11×10^{-2}	62.5
^{15}N	0.365	3.85×10^{-6}	3.65×10^{-3}	935
^{57}Fe	2.19	7.39×10^{-7}	2.19×10^{-2}	29635
^{77}Se	7.85	5.26×10^{-4}	7.58×10^{-2}	144.1
^{113}Cd	12.26	1.34×10^{-3}	12.26×10^{-2}	91.5
^{119}Sn	8.58	4.44×10^{-3}	8.58×10^{-2}	19.3
^{195}Pt	33.8	3.36×10^{-3}	3.38×10^{-1}	100.6
^{199}Hg	16.84	9.54×10^{-4}	1.68×10^{-1}	176.1

a Defined as $R_X^H = |\gamma_X/\gamma_H|^3 \dfrac{N_X I_X(I_X+1)}{N_H I_H(I_H+1)}$, where N is the natural abundance.
b Observation indirectly through ^1H resonance. Defined as $I_X^H = N_X R_H^H$.
c Defined as I_X^H/R_X^H.

Nonproton Spectra

Nuclear magnetic resonance studies of oriented molecules have been confined predominantly to the nuclei H and, to a smaller extent, ^{19}F and ^{31}P. The reasons for this limitation are the low sensitivity of the other nuclei and the fact that the same information can be derived more easily from the ^1H–^nX satellite spectrum.

Proton-coupled ^{13}C spectra may sometimes provide starting parameters for the analysis of the ^1H–^{13}C satellite spectrum. Fluorine-19 and ^{31}P spectra have been investigated mainly for determining anisotropies of the chemical shifts and indirect coupling constants. The analysis of these spectra is similar to that of proton spectra.

For nuclei with $I \geqslant 1$ the Hamiltonian is modified by the addition of a term that takes into account the coupling energy between the nuclear quadrupole moment and the electric field gradient at the site of the nucleus [see eq. (2-33)].

If we define the quadrupole tensor as follows:

$$q_{izz} = \frac{eQ_i V_{zz}}{h} \tag{2-36}$$

then in a molecule's fixed coordinate system (a,b,c) we have

$$\langle q_{izz} \rangle = \frac{2}{3} \sum_{\alpha,\beta}^{a,b,c} S_{\alpha\beta} q_{i\alpha\beta} \tag{2-37}$$

and

$$\hat{H}_Q = \sum_i \sum_{\alpha,\beta} q_{i\alpha\beta} S_{\alpha\beta} [3\hat{I}_{iz}^2 - I_i(I_i + 1)]/6I_i(2I_i - 1) \tag{2-38}$$

Neglecting the constant term and considering only deuterons with $I = 1$, we obtain:

$$\hat{H}_Q = \sum_i B_{ii} \hat{I}_{iz}^2 \tag{2-39}$$

where

$$B_{ii} = \frac{3}{4} \langle q_{izz} \rangle = \frac{1}{2} \sum_{\alpha,\beta} S_{\alpha\beta} q_{i\alpha\beta} \tag{2-40}$$

Equation (2-39) shows that the energy level of a deuteron are equally affected for the states $|1\rangle$ and $|-1\rangle$ but unaffected for $|0\rangle$. Consequently, the frequencies corresponding to the transitions $|1> \rightarrow |0>$ and $|0> \rightarrow |-1>$ differ by $2B_{ii}$.

For molecules with a symmetry of C_3 or higher, eq. (2-40) is simply:

$$B_{ii} = \tfrac{3}{4} q_{cc} S_{cc} \tag{2-41}$$

where the c axis lies along the bond to deuterium and $q_{cc} = q_D$ is the quadrupole coupling constant.

The Hamiltonian of the scalar coupling between chemically equivalent nuclei with $I \geqslant 1$ does not commute with the total spin Hamiltonian. The scalar couplings therefore have influence on the spectrum and the nuclei in question cannot be magnetically equivalent in oriented molecules.[46]

The quadrupole coupling tensor, being traceless, does not affect the transition energies of spectra recorded in isotropic solutions. However the quadrupolar interaction is a very efficient relaxation mechanism. It broadens the lines and makes such nuclei as chlorine, bromine, and iodine very difficult to observe, particularly in anisotropic phases.

From the point of view of line broadening the most favorable nucleus is the deuteron, which has quite small quadrupole coupling constants. In many cases even the D–D direct couplings, which are approximately 40 times smaller than the corresponding H–H couplings, are observable. Also, ^{2}H–^{13}C satellite spectra with ^{13}C natural abundance and ^{2}H enrichment have been recorded recently.[47,48]

Structural Data from Spectra

Direct Coupling and Molecular Structure

A linear molecule of three nuclei (ABC) represents the simplest system the structure of which may be studied from the spectrum of the oriented molecule. The observed couplings are

$$D_{AB} = -\frac{1}{r_{AB}^3} \cdot S_{AB} \cdot K_{AB}$$

$$D_{BC} = -\frac{1}{r_{BC}^3} \cdot S_{BC} \cdot K_{BC} \qquad (2\text{-}42)$$

$$D_{AC} = -\frac{1}{(r_{AB} + r_{BC})^3} \cdot S_{AC} \cdot K_{AC}$$

where $S_{AB} = S_{BC} = S_{AC} = S$. For three protons $K_{AB} = K_{BC} = K_{AC} = \mu_o \hbar \gamma^2{}_H / 8\pi^2$. A list of K values for various nuclei is presented in Table 2-3.

TABLE 2-3. GYROMAGNETIC RATIOS γ AND $\sqrt{K}$ VALUES OF SOME COMMON NUCLEI

Nucleus	γ (10^7 rad T^{-1} s^{-1})	$\sqrt{K}$ (10^{-10} $m^{3/2}$ $s^{-1/2}$)	$\sqrt{K}$ (Å$^{3/2}$ $s^{-1/2}$)
^{1}H	26.7510	10.960	346.556
^{2}H	4.106	1.682	53.190
^{13}C	6.7263	2.756	87.155
^{15}N	−2.7107	−1.111	−35.129
^{19}F	25.1665	10.310	326.024
^{31}P	10.829	4.436	140.279

Each pair of Ds determines the distance ratio that fully describes the molecular structure:

$$D_{AB}/D_{BC} = r_{BC}^3/r_{AB}^3 \qquad (2\text{-}43)$$

Even in this simple system a basic limitation of the method can be seen: only internuclear distance ratios, ie, molecular shapes, can be measured and not absolute lengths. Furthermore, if it is desired to determine the degree of order S, an absolute internuclear distance must be assumed. These limitations are quite general because in all the measured direct couplings, the unknown parameters contribute in the form S/r^3, so that a change in r can be compensated by the corresponding change in the degree of order S.

If a slightly more complex system is examined, four nuclei at the corner of a rectangle, it can be seen that three different direct couplings may be determined. Here for the first time the problem of determinacy must be faced: are there enough measurable direct couplings to define the unknown geometry as well as the order parameters? Choosing the distance between two nuclei as a basis and remembering that only the shape of the rectangle can be derived, it is obvious that one coordinate of the third nucleus is the only unknown structure parameter. Furthermore, there are three unknown and three measured parameters, so the system is exactly determined. In general it can be demonstrated that in order to be determined or overdetermined the number n of nuclei of a system must be equal to or exceed a certain limit, as follows:

1. Three-dimensional system without symmetry, $n \geqslant 7$
2. Two-dimensional system without symmetry, $n \geqslant 5$
3. Two-dimensional system with C_2 symmetry, $n \geqslant 4$

The relations between the internuclear distances and the direct couplings are usually very complex. An example is the equation for a rectangle:

$$D_{12}(r_{12}/r_{23})^5 - D_{13}[1 + (r_{12}/r_{23})^2]^{\frac{5}{2}} + D_{23} = 0 \qquad (2\text{-}44)$$

This equation has no analytical solutions, so that generally iterative computer methods are used for the structure determination.

In contrast, eq. (2-44) is simple enough to study the problem of the uniqueness of solutions. For this particular system it turns out that depending upon the ratios D_{12}/D_{23} and D_{13}/D_{23} the equation has two, one, or zero solutions, of which, of course, only one is physically feasible. As long as the two solutions differ considerably, it is easy to identify the feasible one. If they are similar, however, it may be impossible to make this identification. The two solutions become identical if the two degrees of order also are accidentally equal. The case of zero solutions is very rare but may be observed if near the point of two identical solutions, any of the measured Ds is slightly in error.

For larger spin systems, which usually are overdetermined, the structure solutions should be unique.

Concerning the simple four-spin rectangular system, it also can be seen that errors in structure determination depend in a very complex way upon the measured parameters. They are, eg, particularly large in the vicinity of a degenerate case of two identical solutions.

Experience shows, that for more complex systems it cannot be found out in detail why the structure precision is quite different if the molecule is dissolved in different liquid crystal solvents.

As pointed out, the direct couplings determine molecular shape; obviously this shape includes the absolute determination of angles. An example is the methyl group AX_3 (with $A = {}^{13}C$), for which the following relations are valid:

$$r_{XX}/r_{AX} = \sqrt{3}\,\sin\beta = 2\,\sin(\alpha/2) \tag{2-45}$$

with $\alpha = \sphericalangle(X, A, X)$ and $\beta = \sphericalangle(C_3 \text{ axis}, X, A)$ and

$$D_{AX}/D_{XX} = \frac{\gamma_A}{\gamma_X} \cdot \left(\frac{r_{XX}}{r_{AX}}\right)^3 \left[\left(\frac{r_{XX}}{r_{AX}}\right)^2 - 2\right] \tag{2-46}$$

The overdeterminacy of a problem may be considerably increased if the information provided by ^{13}C satellites in the spectra of the oriented proton system is included.

In the case of the rectangle of four spins, with one ^{13}C included the number of observed couplings increases from three to seven. At the same time the number of unknowns is increased by the two ^{13}C coordinates, ie, from three to five. The system is now overdetermined by two equations. For systems with less symmetry and consequently with a large variety of ^{13}C positions the overdeterminacy increases even more drastically.

Computer Program for Structural Determination

It has been seen that even for small spin systems the use of a computer is necessary to determine molecular structure. Various programs have been written for this purpose. Here, only one that is actually used widely is discussed; it is the only one that also exists in a PASCAL version, which may be run on a normal spectrometer-dedicated computer, as, eg, the ASPECT 2000.

The program SHAPE[49] uses an estimated starting structure, estimated order parameters and known symmetry elements, and a fixed internuclear distance (as a basis), together with the measured D couplings with errors as input. It compares iteratively the Ds of a computer structure with the measured Ds, varying the structure, as well as the order parameters, until the rms error of the fit is minimized. It then prints out nuclear coordinates, internuclear distance ratios, absolute angles, and order parameters, as well as the probable errors of all the parameters based on the variance–covariance matrix of the measured Ds.

The program allows for inclusion of intramolecular motion (see The Study of Intramolecular Motion).

It is also possible to iterate on the information of several spectra at the same time, ie, to determine one molecular geometry and several sets of order parameters.

The Study of Intramolecular Motion

This section explains how intramolecular motion may be considered in the structural analysis performed by the program SHAPE. The various problems and limitations of the method are discussed in the sections Molecular Vibrations and Correlation.

Internal motion can be included in the SHAPE program as follows:

A hindering potential is chosen of the form:

$$V = \sum_{n=1}^{k} \frac{V_n}{2}(1 - \cos n\phi) \tag{2-47}$$

The probability distribution of the rotor is then either assumed classically, as

$$P(\phi)\, d\phi = \frac{\exp(-V/kT)\, d\phi}{\int_0^{2\pi} \exp(-V/kT)\, d\phi} \tag{2-48}$$

with

$$D = \int_0^{2\pi} D(k) P(\phi)\, d\phi \tag{2-49}$$

or, alternatively, the quantum mechanical rotor is considered and the corresponding $P(\phi)$ computed. Actually for methyl groups the results turn out to be similar as long as $V < 10$ kJ mol^{-1}.

Generally, the sensitivity of the direct coupling to the potential decreases rapidly with increasing rotational symmetry. For example $D = f(\phi)$ of the coupling between a rotating methyl group and the ortho ring proton varies by 500 Hz for a onefold, by 100 Hz for a threefold, and by 10 Hz for a sixfold symmetry (for an assumed fixed degree of order).

Precision of Structural Results

During the last 20 years at least 500 different molecular structures have been derived by NMR of the oriented molecule. Large tables of the resulting data have been collected and may be found in review articles[8-18] and books.[6,7]

In principle, the theoretically achievable precision is extremely high. Direct coupling between protons of the order of 10^3 Hz can be measured to 10^{-1} Hz so that the relative precision in Ds can be as large as 10^{-4} and, because the Ds are proportional to r^{-3} the maximum distance precision $\Delta r/r$ may be 10^{-5}. In practice, however, this precision is usually meaningless. Vibration corrections and particularly correlation effects introduce errors that reduce the distance precision to 10^{-3} or even 10^{-2}. Typically errors in C—H bond lengths are 10^{-2} Å and because of the reduction of precision on going from H–H to C–H couplings the C—C bond lengths are even less precise with errors of several 10^{-2} Å. The NMR method of oriented molecules seems particularly precise for angle measurements, for which precisions of a few seconds can be obtained. However, as long as the solvent effects on the structure are not understood fully as yet, the precision may be only apparent and the actual precision may be of the order of 10^{-1} degrees.

For molecules whose shape is defined by symmetry, such as benzene, and for which the solvent effects can be corrected, however, the precision that may be attained is very high. For instance the distance ratio $r_z(CH)/r_z(CC)$ was derived[50] as 0.77327 ± 0.00007, which agrees with infrared (IR) and electron diffraction (ED) results (0.777 ± 0.004) but is one hundred times more precise.

Limitations to the Precision of Structure Determination

Anisotropy of Indirect Coupling

As pointed out in the section, Basic Theory, the general coupling tensor consists of two contributions, indirect and direct. In isotropic solutions the trace of the indirect coupling tensor is measured as the "indirect coupling constant". The trace of the direct coupling tensor is zero. In anisotropic solutions the averages of both tensors can be observed. Unfortunately, they have a similar orientation dependence, so that the average is always the sum of the two types. Only in overdetermined systems, which allow, eg, the measurement of a nuclear position simultaneously and independently by two different couplings, is it possible to separate direct and indirect contributions.

In practice, for many pairs of nuclei, particularly if one member of the pair is a proton, it is usually assumed that the anisotropy of the indirect coupling may be neglected. There are theoretical as well as experimental confirmations for such an approach.

An extensive list of anisotropies of J couplings may be found in a review article by Lounila and Jokisaari.[31] The orders of magnitude of the effects may be summarized as follows: for couplings between nondirectly bonded protons the anisotropy of J should be zero. For H–C and H–N the predicted and measured effects ($|J_{ij}^{aniso}/2D_{ij}|$) are smaller than 0.1%. For heavier nuclei X in X—H bonds the anisotropy may become significant. Finally for such pairs of nuclei as F–F, F–C, F–P, and C–C the anisotropy may reach one or even several percent. If the internuclear axis happens to lie near the direction for which the degree of order vanishes (see the section, The Order Tensor) the ratio $|J_{ij}^{aniso}/2D_{ij}|$ may become very large; because J_{ij}^{aniso} is not generally zero for zero degree of order.

It is clear that the neglect of anisotropy, particularly of the C—H bonds, introduces small but systematic errors into structures as determined by NMR of oriented molecules. Furthermore, the measured coupling between two carbon atoms should not be used for the structure determination. For this pair of nuclei the neglected anisotropy generally is larger than the measurement error.

In principle, experiments that determine the anisotropy of J can be performed. However, in the light of recent results on solvent effects on structure,[51,52] one should avoid combining data for a molecule in several different solvents, particularly if the anisotropy is small.

Molecular Vibrations

The NMR of oriented molecules can measure, for example, relative bond lengths. If it is remembered the internuclear distance of approximately 1.1 Å for a C—H bond and considering that because of IR vibrations this distance varies by roughly ± 0.1 Å in time and, furthermore, that also the degree of order of the axis is time dependent, the question arises: what is a bond length? Obviously an average value must be defined and will be different for various spectroscopies. In electron diffraction $\langle r \rangle$, is measured with microwaves $\langle r^{-2} \rangle^{-\frac{1}{2}}$ and with NMR approximately $\langle r^{-3} \rangle^{-\frac{1}{3}}$. In order to compare NMR results with data of different methods, the measured direct couplings must be corrected:

$$D = -\mathrm{K} \sum_{\alpha, \beta} S_{\alpha\beta} \phi_{\alpha\beta} \tag{2-50}$$

where

$$\phi_{\alpha\beta} = \langle r_\alpha r_\beta / r^5 \rangle \tag{2-51}$$

If we assume that S does not vary as a function of the vibration $\phi_{\alpha\beta}$ may be expanded for small amplitudes of vibration (Δ_μ). The resulting terms:

$$\phi_{\alpha\beta} = \phi_{\alpha\beta}^{\mathrm{e}} + \phi_{\alpha\beta}^{\mathrm{a}} + \phi_{\alpha\beta}^{\mathrm{h}} + \ldots \tag{2-52}$$

correspond to the equilibrium geometry, the anharmonic correction, and the harmonic correction. Usually only the harmonic correction is performed because the information on the anharmonic potentials is very scarce. The observed direct coupling is a superposition:

$$D = D^{\mathrm{e}} + d^{\mathrm{a}} + d^{\mathrm{h}} \tag{2-53}$$

After calculation of the harmonic correction, which must be based on a corresponding force field and is derived by such computer programs as VIBR[52] a so-called r_z structure is obtained:

$$D^{z} = D - d^{\mathrm{h}} = D^{\mathrm{e}} + d^{\mathrm{a}} \tag{2-54}$$

This structure corresponds to the average nuclear positions and has the advantage of being additive (no shrinkage effects).

For simple molecules the results can be given in analytical form. For nuclei on a threefold or higher axis of symmetry (z axis):

$$d^{\mathrm{h}} = 6(C_{zz} - C_{xx}) D^{\mathrm{e}} / R^2 \tag{2-55}$$

with R = internuclear distance and $C_{\alpha\beta}$ = covariance matrix of vibrations = $\langle \Delta_\alpha \Delta_\beta \rangle$. The corresponding anharmonic term is

$$d^{\mathrm{a}} = -3 \langle \Delta z \rangle D^{\mathrm{e}} / R \tag{2-56}$$

For a system of nuclei (on the z axis) with two planes of symmetry:

$$d^{h} = D^{e}\frac{1}{R^{2}}\left[\frac{1}{2}(12C_{zz} - 7C_{xx} - 5C_{yy}) + (C_{yy} - C_{xx})(S_{yy}/S_{zz})\right] \quad (2\text{-}57)$$

In general the harmonic corrections decrease with the square of the internuclear distance. They are particularly large if one degree of order (S_{zz}) approaches zero (see eq. 2-57). In such cases a percentage correction usually fails.

Taking benzene as an example (with $S_{xx} = S_{zz} = -\frac{1}{2}S_{yy}$):

$$d^{h} = D^{e}\frac{1}{R^{2}}\left[\frac{1}{2}(12C_{zz} - 9C_{yy} - 3C_{xx})\right] \quad (2\text{-}58)$$

For a C—H bond, for example, $C_{xx} = 1.41 \times 10^{-2}$ Å^2, $C_{yy} = 2.26 \times 10^{-2}$ Å^2, and $C_{zz} = 0.59 \times 10^{-2}$ Å^2. The correction d^h is $-0.072\, D^e$; i.e., there is a considerable vibration correction of 7%, leading to an error in the distance of approximately 2×10^{-2} Å if neglected. For benzene the contributions to the corrections predominantly stem from C—H in and out of plane deformations.

In the derivation of vibration corrections as discussed so far, it has been assumed that the molecular orientation is independent of the vibrational motion. However, if the ordering potential depends upon the vibrational state, then:

$$D = K \sum_{\alpha,\beta,v} P_{v}\phi^{v}_{\alpha\beta}S^{v}_{\alpha\beta} \quad (2\text{-}59)$$

with $P_v = \exp(-U^{v}_{int}/kT)/\sum_v \exp(-U^{v}_{int}/kT)$; ie, each vibrational mode contributes to the coupling with its separate order tensor. It can only be hoped that this is not usually the case; else, because of the large number of parameters, spectral analysis would become very difficult or impossible. On the other hand, for very small and highly symmetrical molecules, such as methane, this approach has been suggested as used for explaining the observable finite direct coupling that actually should be zero.[53]

Correlation

In the section, Basic Theory it was seen that the direct coupling was an average of a product of two tensors, one of which described the orientation, the other the geometry of the spin system. As long as orientation and geometry are independent of each other, the two may be separated and individual averages can be determined. If, however, the two factors are interdependent, ie, if there is correlation, this separation is no longer correct. The product must be averaged as a whole. This phenomenon of correlation has several consequences which may be described in a simple model as follows:

Assuming that the internuclear distance depends upon the angle Θ between the internuclear axis and the liquid crystal optic axis,

$$r = r_{o}[1 + f(\Theta)] \quad (2\text{-}60)$$

with $f(\Theta) \ll 1$, then

$$r^{-3} = r_o^{-3}[1 - 3f(\Theta)] \tag{2-61}$$

and the direct coupling is

$$\begin{aligned} D' &= -K\, r_o^{-3}\{\langle P(\cos\Theta)[1-3f(\Theta)]\rangle\} \\ &= -K\, r_o^{-3}[S - 3\langle P_2(\cos\Theta)f(\Theta)\rangle] \end{aligned} \tag{2-62}$$

There is now a normal term $-KSr_o^{-3}$ plus a correlation term which must be discussed further.

The function $f(\Theta)$ is apolar with a period of π; it is an even function; therefore:

$$f(\Theta) = \sum_{k=0}^{\infty} f_k[\sin(2\Theta), \cos(2\Theta)] \tag{2-63}$$

In its simplest form

$$f(\Theta) = f_o + f_1 \cos(2\Theta)$$

where $f_o = -\alpha$ and $f(0) = 0$. Then:

$$f(\Theta) = \alpha[\cos(2\Theta) - 1] \tag{2-64}$$

and

$$r = r_o\{1 + \alpha[\cos(2\Theta) - 1]\} \tag{2-65}$$

The resulting direct coupling D' is now:

$$D' = -Kr_0^{-3}\left[\left(1 + \frac{20}{7}\alpha\right)S - \frac{4}{5}\alpha - \frac{72}{35}\alpha\langle P_4(\cos\Theta)\rangle + \text{higher terms}\right] \tag{2-66}$$

It contains contributions proportional to S (normal term) and proportional to αS (mixed term). Of particular interest is a term proportional to α, a pure correlation term. This term is small, being the maximum relative change of a bond length ($\Delta r/r$) or bond angle ($\Delta\Phi/\Phi$), but if it is not reduced with decreasing degree of order, it becomes important for small S.

With $D(\text{correlation})/D = \alpha/S$ it can be seen that for $\alpha = 10^{-3}$ and $S = 10^{-2}$, ie, $\alpha/S = 10^{-1}$, there is already a 10% error in a D ratio, or a 3% error in a distance ratio, if correlation is neglected.

Correlation is important not only for small degrees of order but quite generally, if a high precision of a measurement is required. An example is the molecule benzene.[51] Obviously this molecule must have a hexagonal structure if the vibration corrections have been performed. In fact there still are some discrepancies observed. For the H—H distance ratio, r_p/r_o, the value is found between 1.993 and 2.004 and for r_m/r_o between 1.725 and 1.734 (errors ± 0.001) depending upon the liquid crystal solvent. For a hexagon the values should be 2.000 and 1.372, respectively.

The deviations can be interpreted as correlation effects.[51] The C—H bonds are assumed bent out of the plane into the direction of the liquid crystal optical axis. A maximum bend angle of 0.1° accounts for the observed effects and explains the deviations, not only of the three H—H, but also of the four C–H direct couplings with only two parameters fitted, the bend angle and the liquid crystal degree of order.

Intramolecular Motion

A case of particularly large correlation effects is intramolecular motion. Obviously there is considerable variation of geometry, which may lead to appreciable variation of the order tensor.

In early studies of intramolecular motion by NMR of oriented molecules[54] it was pointed out that there were various lifetimes in this field that affected the observable spectra, as well as their analysis. If a molecule, for example, remains in a particular conformer longer than the inverse difference between direct couplings in its various conformers, several spectra are observed, one for each conformer. If the lifetime is shorter, an average spectrum is observed that may be difficult to interpret for several reasons: first, it turns out that with all the measured direct couplings only products of conformer concentration and order parameters can be determined. Second (eg, for a case of two conformations), the order parameters may depend upon the lifetimes in a complex way, as follows:[55]

$$S_1 = S_{1\infty} + \frac{(S_{2\infty} - S_{1\infty})k_2\tau_2}{2\tau_1\tau_2k_1k_2 + k_1\tau_2 + k_2\tau_2} \tag{2-67}$$

$$S_2 = S_{2\infty} + \frac{(S_{1\infty} - S_{2\infty})k_1\tau_1}{2\tau_1\tau_2k_1k_2 + k_1\tau_1 + k_2\tau_2} \tag{2-68}$$

with k the rate constant for molecular reorientation, τ the lifetime of conformer, and $S_{\nu\infty}$ the order parameter of stable conformer ν. For $\tau \ll k^{-1}$ an average order tensor is observed:

$$S_1 = S_2 = \frac{k_1\tau_1S_{1\infty} + k_2\tau_2S_{2\infty}}{k_1\tau_1 + k_2\tau_2} \tag{2-69}$$

In a second, more recent analysis an equilibrium statistical mechanical approach to the problem[56] has been suggested. It shows that quite generally the observed direct couplings are averages:

$$D = -K \sum_{\alpha,\beta} \sum_{\nu} P_\nu S^\nu_{\alpha\beta} \phi^\nu_{\alpha\beta} \tag{2-70}$$

with P_ν the concentration of conformer ν; α,β the x,y,z internal coordinates, and $S^\nu_{\alpha\beta} = \langle \frac{1}{2}(3l^\nu{}_\alpha l^\nu{}_\beta - \delta_{\alpha\beta})\rangle$,
where l^ν_β and l^ν_β are direction cosines in molecular frame; $\delta_{\alpha\beta}$ is the Kronecker delta symbol, and:

$$\phi^\nu_{\alpha\beta} = \left\langle \frac{r^\nu_\alpha r^\nu_\beta}{(r^\nu)^5} \right\rangle \tag{2-71}$$

If the external energy of the conformers is independent of the internal coordinates (ie, of the conformation), eg, (2-70) simplifies to:

$$D = -K \sum_{\alpha,\beta} \sum_{\nu} S_{\alpha\beta} P_{\nu} \phi_{\alpha\beta}^{\nu} \tag{2-72}$$

with an "average" degree of order. Actually to be more precise, the sum over conformers should be replaced by an integral over internal coordinates.

In general, there will be a potential barrier to intramolecular motion that must be characterized by its amplitude, its shape, and the positions of the minima. Usually a simple cosine potential is chosen and the corresponding probabilities may be determined, either quantum mechanically by solution of the Mathieu-type equation of the rotor, or from a classical point of view with continuous Boltzmann distribution. At any rate, the potential introduces at least two new unknowns (amplitude and position of the barrier) into the analysis. Consequently, two or more direct couplings should be measurable that are sensitive to the motion. It should also be mentioned that the symmetry of the potential has a drastic effect upon the sensitivity with a decrease from two- to three- to sixfold axes.

In early studies predominantly methyl group rotations were analyzed; later on, aldehyde group rotations were examined also and, finally, large conformational changes. In many of these studies quite drastic simplifying assumptions were made, for example, that there existed an average order tensor or that the conformers existed exclusively at the minium of the potential curve. These early studies, except for the methyl rotation, now seem to be questionable in the light of more recent approaches, which show that actually the degree of order varies continuously with the conformational motion. The exact prediction of this variation, of course, is quite difficult, but the results of the theoretical approach can be shown to be qualitatively predictable and understandable. If the puckering motion of cyclopentene is taken as an example[57] with the C_2 axis being the x axis, the second axis of the molecular plane being the y axis, and the z axis out of the plane, it can be seen that the motion should not affect S_{yy}, because the y dimension of the molecule is unaffected. Also, S_{xx} should decrease simultaneously with the puckering angle, because the molecular diameter increases in the z and decreases in the x direction. Finally, S_{xz} should go through a maximum at 45° because it must be zero at 0° and 90°.

A last point to be mentioned is a quite general one but particularly important in studies of intramolecular motion: this is the problem of the "exactly determined case". Spectral parameters derived in an analysis usually highly overdetermine the structural problem and consequently the rms error of the least-squares fit is used as criterion for the correctness of the analysis. This situation is no longer ideal if not only the molecular geometry, but also the intramolecular motion, must be derived from the measured direct couplings. Quite often such problems are only slightly overdetermined or exactly determined. By "slightly overdetermined" it is meant that there are more equations than unknown parameters, but that some of them are insensitive to one or more parameters. In such fits the rms error is no longer a criterion. It should be zero, anyway, for the exactly determined case. If, for example, one wants to determine the angle between benzene planes in substituted biphenyls and has only one measurable inter-ring coupling, one will of course obtain an extremely precise angle, which, unfortunately, is not unique, although the rms error of the fit is zero. Only if one is able to measure many such couplings can one study details of the potential, parameters that characterize the real system "behind" the measured couplings and not a hypothetical stable configuration.

In summarizing the dangers of the exactly determined case, it can be stated the rms error, which is small or zero in slightly overdetermined systems, does not confirm the correctness of a model for the internal motion, but only means that the derived molecular parameters are in agreement with both the assumed non-unique model and the measured couplings.

Multiple Sites

A special type of correlation may be observed if a solute molecule interacts at specific sites with the solvent liquid crystal and therefore exists in at least two forms, the free molecule and the complex. Such a molecule has at least two degrees of order, S_1 and S_2, for the solute, which may vary slightly its geometry g in each site.

As an example, consider a linear molecule of three spins with two possible sites:

$$\left.\begin{aligned} g_{A2} &= g_{A1}\ (1+\beta_A),\ \beta_A \ll 1 \\ g_{B2} &= g_{B1}\ (1+\beta_B),\ \beta_B \ll 1 \end{aligned}\right\} \quad (2\text{-}73)$$

$$\begin{aligned} D_A &= pS_1g_{A1} + (1-p)S_2g_{A2} = g_{A1}[pS_1 + (1-p)S_2(1+\beta_A)] \\ D_B &= pS_1g_{B1} + (1-p)S_2g_{B2} = g_{B1}[pS_1 + (1-p)S_2(1+\beta_B)] \end{aligned} \quad (2\text{-}74)$$

with p_v the relative concentration of the solute at site v. D_A and D_B are couplings between nuclei on a common axis and should therefore, normally, be equal to zero at the same time, ie, if the degree of order of the axis is equal to zero. However, because of correlation, D_B is finite for $D_A = 0$:

$$D_B(D_A=0) = g_{B1}\ [(1-p)S_2(\beta_B - \beta_A)] \quad (2\text{-}75)$$

and:

$$D_A(D_B=0) = g_{A1}\ [(1-p)S_2(\beta_A - \beta_B)] \quad (2\text{-}76)$$

A necessary condition for the possibility that one D value is zero is:

$$pS_1 = -(1-p)S_2\ (1+\beta_B)$$

ie, S_1 and S_2 must have opposite signs.

As a consequence of correlation, distance ratios, which are defined by D_A/D_B and should be constant, are now variables and may even assume the values zero and infinity.

It is important to note that a quite small β may induce rather large apparent variation of the geometry, because the zero point or pole of the D_A/D_B ratio affects the function to a considerable distance from the singularity.

A practical example of the multisite behavior was first observed for the molecule acetylene,[58] which in one liquid crystal displayed an apparent structure variation of 30% as a function of temperature. The anomaly could be explained by the existence of complex formation with opposite sign of order and a true variation of molecular shape of 1%.

The multiple site situation can also be realized artificially by using two liquid crystals that orient the solute with opposite signs of the degree of order. As an example methylalcohole[59] has been demonstrated to show an apparent variation of the methyl angle of several degrees with a true variation of 0.3°, if dissolved in the liquid crystal mixture.

Solvent Effects

Various types of solvent effects on the structures of oriented molecules have been observed. Generally the HCH angles of methyl groups seem to be solvent dependent with a variation of the order of 1°. Bond lengths also have been found variable, particularly C—C bonds and to a slighter extent C—H bonds. The variations may be of the order of several hundredths of an Ångstrom.

These considerable variations are difficult to understand. Simple models in general predict smaller effects. For the methyl group, for example, it can be shown[60] that the deformation of an XCH angle by dispersion interaction should increase with degree of order. In molecules of the series CH_3I, CH_3Br, CH_3Cl, CH_3F, and for $S=0.3$, the deformation should increase from a few thousandths of a degree to a few hundredths of a degree.

Another model, in which the interaction of the solute with the solvent is simulated by an electric charge that approaches a molecule on the C_3 axis, predicts no measurable change of structure for one positive electron charge at 10-Å distance from the methylcarbon of methylcyanide but a decrease of the CCH angle by approximately 1° for a distance of 3 Å. At the same time the C–N distance is increased from 1.14 to 1.44 Å. Obviously, in practice the effects must be much smaller because the solvent local charges will be only a small fraction of an electron.

One should not forget, however, that because of correlation effects combined with sites of opposite sign for the degree of order, minute actual solvent effects on molecular shape may be considerably amplified into apparently large effects (see the section Multiple Sites).

Conclusions

Since the invention of NMR spectroscopy of molecules dissolved in liquid crystals, approximately 500 different molecules have been investigated. These studies, however, do not only deal with solute molecular structures but many of them concentrate on S tensors, chemical shift anisotropies, spin–spin coupling anisotropies, or quadrupole coupling constants. Mostly ^{1}H, ^{2}H, ^{19}F, and ^{31}P are used, but also ^{11}B, ^{13}C, ^{14}N, ^{15}N, ^{17}O, ^{111}Cd, ^{113}Cd, ^{117}Sn, ^{119}Sn, and ^{199}Hg resonances have been detected in thermotropic liquid crystals. Furthermore, the following nuclei have been observed as ions in lyotropic liquid crystals:[18] ^{7}Li, ^{23}Na, ^{35}Cl, ^{81}Br, ^{85}Rb, and ^{135}Cs.

Relatively large molecules have been studied by using ^{1}H NMR spectra exclusively. However, complete structures (including, for example, carbon positions) cannot be determined for very large molecules. In practice, molecules the size of substituted benzene, such as *p*-chlorotoluene[61] and benzaldehyde,[62] form the upper limit. This is because the complexity of a spectrum increases very rapidly with increasing number of spins. For example, up to 1133 lines could be assigned to *p*-chlorotoluene in Merck ZLI 1132. Furthermore, the inclusion of D_{CH} coupling constants in structure determinations presumes knowledge of the harmonic molecular force fields, which is necessary to perform vibration corrections to coupling constants.

Concerning the future development of NMR with oriented molecules as used for structure determination it may be concluded that the most important object of research at the moment is solvent effects: by studying systematically small and highly symmetric molecules, using various liquid crystals as solvents, it may be possible to find the reasons for the observed variation of solute structure as a function of solvent. Obviously theoretical studies are also required. A first attempt in this direction has been made for methane,[53] but the model must be developed for more complex systems.

Variation of solute molecular structure may be an actual permanent deformation, perhaps because of complex formation, or it may be an apparent deformation caused by correlation, ie, coupling between vibration and rotation. Only if it becomes possible to distinguish between these alternatives, and also to understand them, will the NMR of oriented molecules be regarded as a fast, precise, and reliable method for determining anisotropic properties of NMR parameters and, in particular, solute molecular structure.

References

1. Buckingham, A. D.; Pople, J. A. *Trans. Faraday Soc.* **1963**, *59*, 2421.
2. Hilbers, C. W.; MacLean, C. In "NMR Basic Principles and Progress", Diehl, P.; Fluck, E.; Kosfeld, R., Eds.; Springer-Verlag: Heidelberg, 1972; Vol. 7, p. 3.
3. Ducros, P. *Bull. Soc. Franc. Mineral. Crist.* **1960**, *83*, 85.
4. Saupe, A.; Englert, G. *Phys. Rev. Lett.* **1963**, *11*, 462.
5. Englert, G.; Saupe, A. *Z. Naturforsch.* **1964**, *19a*, 172.
6. Diehl, P.; Khetrapal, C. L. In "NMR Basic Principles and Progress", Diehl, P.; Fluck, E.; Kosfeld, R., Eds.; Springer-Verlag: Heidelberg, 1969; Vol. 1, p. 3.
7. Emsley, J. W.; Lindon, J. C. "NMR Spectroscopy Using Liquid Crystal Solvents"; Pergamon: Oxford, 1975.
8. Diehl, P.; Henrichs, P. M. In "NMR Specialist Periodical Reports", Harris, R. K., Ed.; The Chemical Society London, 1972; Vol. 1, p. 321.
9. Bulthuis, J.; Hilbers, C. W.; MacLean, C. *MTP Int. Rev. Sci. Phys. Chem., Ser. 1* **1972**, *4*, 201.

10. Diehl, P.; Niederberger, W. In "NMR Specialist Periodical Reports", Harris, R. K., Ed.; The Chemical Society London, 1974; Vol. 3, p. 368.
11. Snyder, L. C.; Meiboom, S. In "Critical Evaluation of Chemical and Physical Structural Information", Lide, D. R.: Paul, M. A., Eds.; National Academy of Sciences: Washington, D.C., 1974.
12. Khetrapal, C. L.; Kunwar, A. C.; Tracey, A. S.; Diehl, P. In "NMR Basic Principles and Progress", Diehl, P.; Fluck, E.; Kosfeld, R., Eds.; Springer-Verlag: Heidelberg, 1975, Vol. 9, p. 3.
13. Lunazzi, L. In "Determination of Organic Structures by Physical Methods", Nachod, F. C.; Zuckermann, J. J.; Randall, E. W., Eds.; Academic: New York, 1976; Vol. 6, p. 335.
14. Diehl, P. In "NMR Specialist Periodical Reports", Harris, R. K., Ed.; The Chemical Society London, 1976; Vol. 5, p. 314.
15. Khetrapal, C. L.; Kunwar, A. C. In "Advances in Magnetic Resonance", Waugh, J. S., Ed.; Academic: London, 1977; Vol. 9.
16. Khetrapal, C. L.; Kunwar, A. C. In "NMR Specialist Periodical Reports", Abraham, R. J., Ed.; The Chemical Society London, 1979; Vol. 8.
17. Khetrapal, C. L.; Kunwar, A. C. In "NMR Specialist Periodical Reports", Abraham, R. J., Ed.; The Chemical Society London, 1980; Vol. 9.
18. Schumann, C. In "Handbook of Liquid Crystals", Kelker, H.; Hatz, R., Eds.; Verlag Chemie: Weinheim, 1980; p. 426.
19. Gray, G. W. "Molecular Structure and Properties of Liquid Crystals", Academic: New York, 1962.
20. Maier, W.; Saupe, A. *Z. Naturforsch.* **1958**, *A13*, 564; **1959**, *A14*, 882; **1960**, *A15*, 287.
21. Jokisaari, J.; Väänänen, T.; Lounila, J. *Mol Phys.* **1982**, *45*, 141.
22. Saupe, A. *Z. Naturforsch.* **1964**, *A19*, 161.
23. Snyder, L. C. *J. Chem. Phys.* **1965**, *43*, 4041.
24. Canet, D.; Marchal, J. P.; Sarteaux, J. P. *C. R. Acad. Sci. Paris*, **1974**, *279*, 71.
25. Diehl, P. In "Nuclear Magnetic Resonance Spectroscopy of Nuclei Other than Protons", Axenrod, T.; Webb, G. A., Eds.; Wiley: New York, 1974; p. 377.
26. Jokisaari, J.; Kuonanoja, J. *JEOL News*, **1977**, *A14*, 8.
27. Jokisaari, J.; Räisänen, K. *Mol. Phys.* **1978**, *36*, 113.
28. Jokisaari, J.; Räisänen, K.; Lajunen, L.; Passoja, A.; Pyykkö, P., *J. Magn. Reson.* **1978**, *31*, 121.
29. Jokisaari, J.; Diehl, P. *Org. Magn. Reson.* **1980**, *13*, 359.
30. Räisänen, K. *J. Mol. Struct.* **1980**, *69*, 89.
31. Lounila, J.; Jokisaari, J. In "Progress in N.M.R. Spectroscopy", Emsley, J. W.; Feeney, J.; Sutcliffe, L. H., Eds.; Pergamon: Oxford, 1982; Vol. 15, part 3.
32. Kelker, H.; Hatz, R. "Handbook of Liquid Crystals"; Verlag Chemie: Weinheim, 1980, p. 465.
33. Ferretti, J. A.; Harris, R. K.; Johannesen, R. B. *J. Magn. Reson.* **1970**, *3*, 84.
34. Diehl, P.; Kellerhals, H. P.; Niederberger, W. *J. Magn. Reson.*, **1971**, *4*, 53.
35. Castellano, S.; Bothner-By, A. A. *J. Chem. Phys.* **1969**, *41*, 3863.
36. Vogt, J. *Bruker Rep.* **1979**, part 3, p. 23.
37. For the use of computers in the analysis of NMR spectra see: Diehl, P.; Kellerhals, H.; Lustig, E. In "NMR Basic Principles and Progress", Diehl, P.; Fluck, E.; Kosfeld, R., Eds.; Springer-Verlag: Heidelberg, 1972; Vol. 6.
38. Diehl, P.; Sýkora, S.; Vogt, J. *J. Magn. Reson*, **1975**, *19*, 67.
39. Diehl, P.; Vogt, J. *Org. Magn. Reson.* **1976**, *8*, 638.
40. Stephenson, D. S.; Binsch, G. *Org. Magn. Reson.* **1980**, *14*, 226.
41. Meiboom, S.; Hewitt, R. C.; Snyder, L. C. *Pure Appl. Chem.*, **1972**, *32*, 251.
42. Lounila, J. and Väänänen, T. *Mol. Phys.* **1983**, *49*, 859.
43. Diehl, P.; Moia, F.; Bösiger, H. *J. Magn. Reson.* **1980**, *41*, 336.
44. Dombi, G.; Amrein, J.; Diehl, P. *Org. Magn. Reson.* **1980**, *13*, 224.
45. Diehl, P.; Dombi, G.; Bösiger, H. *Org. Magn. Reson.* **1980**, *14*, 280.
46. Saupe, A.; Nehring, J. *Chem. Phys.* **1967**, *47*, 5459.
47. Wooten, J. B.; Beyerlein, A. L.; Jacobus, J.; Savitsky, G. B. *J. Magn. Reson.* **1978**, *31*, 347.
48. Jokisaari, J.; Diehl, P. Amrein, J.; Ijäs, E. *J. Magn. Reson.* **1983**, *52*, 193.
49. Diehl, P.; Henrichs, P. M.; Neiderberger, W. *Mol. Phys.* **1971**, *20*, 139.
50. Diehl, P.; Bösiger, H.; Zimmermann, H. *J. Magn. Reson.* **1979**, *33*, 113.
51. Diehl, P.; Jokisaari, J.; Moia, F. *J. Mol. Struct.* **1982**, *96*, 107.
52. Sykora, S.; Vogt, J.; Bösiger, H.; Diehl, P. *J. Magn. Reson.* **1979**, *36*, 53.

53. Emsley, J. W.; Luckhurst, G. R., *Mol. Phys.* **1980**, *41*, 19.
54. Diehl, P.; Henrichs, P. M.; Niederberger, W. *Org. Magn. Reson.* **1971**, *3*, 243.
55. Burnell, E. E.; de Lange, C. A. *J. Magn. Reson.* **1980**, *39*, 461.
56. Emsley, J. W.; Luckhurst, G. R. *Mol. Phys.*, **1980**, *41*, 19.
57. Stephenson, D. S.; Binsch, G. *Mol. Phys.* **1981**, *43*, 697.
58. Diehl, P.; Sykora, S.; Niederberger, W.; Burnell, E. E. *J. Magn. Reson.* **1974**, *14*, 260.
59. Diehl, P.; Reinhold, M.; Tracey, A. S.; Wullschleger, E. *Mol. Phys.* **1975**, *30*, 1781.
60. Lounila, J., Diehl, P., Hiltunen, Y., Jokisaari, J. *J. Magn. Reson.* **1985**, *61*, 272.
61. Diehl, P. Moia, F. *Org. Magn. Reson.*, **1981**, *15*, 326.
62. Diehl, P.; Jokisaari, J.; Amrein, J. *Org. Magn. Reson.* **1980**, *13*, 451.

3

CARBON-13 NUCLEAR SPIN RELAXATION STUDY AS AN AID TO ANALYSIS OF CHAIN DYNAMICS AND CONFORMATION OF MACROMOLECULES

Ryozo Kitamaru

INSTITUTE FOR CHEMICAL RESEARCH, KYOTO UNIVERSITY,
UJI, KYOTO 611, JAPAN

Introduction

When subjected to a static magnetic field, substances, including nuclei with nonzero spin angular momentum, such as ^{1}H and ^{13}C, exhibit equilibrium magnetization of each nucleus in the direction parallel to the magnetic field, according to Curie's law. Let M_x, M_y, and M_z be the magnetization in the x, y, and z directions in a rectangular coordinate whose z axis is parallel to the static field H_0; then their equilibrium values M_x^0, M_y^0, and M_z^0 are given as:

$$M_z^0 = N\gamma^2\hbar^2 H_0 I(I+1)/3kT$$

$$M_x^0 = 0, \; M_y^0 = 0 \qquad (3\text{-}1)$$

Where N denotes the number of spins, γ the magnetogyric ratio, $\hbar$ is Planck's constant/2π, k is Boltzmann's constant, and I is the maximum eigenvalue of the one-dimensional component of the spin operator $\mathbf{I}$ (I is an integer or half integer). Such a magnetization arises as a result of equilibrium between the

NMR in Stereochemical Analysis

interaction energy of the spins with the magnetic field and other degrees of freedom related to the spin system (such as rotation and vibration of atoms and molecules); these degrees of freedom are usually referred to as a lattice. Because the heat content of the spin system is negligibly small in comparison with that of the lattice, the lattice acts as a heat reservoir of a temperature T to the system.

If any appropriate perturbation is added to the spin system in addition to the static field H_0, the magnetization deviates from the equilibrium value. After the perturbation is removed, however, the magnetization tends to return to the equilibrium value. In most cases it has been confirmed experimentally that this relaxation process follows the relation:

$$-\frac{dM_i}{dt}=\frac{M_i-M_i^0}{T_j} \tag{3-2}$$

with $i=x$, y, or z. Here, T_j is a time constant to determine the rate of the relaxation, and it is customarily designated as T_1 for the change of the magnetization in the direction of H_0, ie, the z direction, and as T_2 in directions perpendicular to H_0, such as the x or y directions.

Equation (3-2) can be easily integrated in the form:

$$M_i-M_i^0=(\text{constant})\times\exp(-t/T_j).$$

with terminal conditions: $M_z(\infty)=M_z^0$, $M_x(\infty)=M_x^0=0$:

$$M_z-M_z^0=[M_z(0)-M_z^0]e^{-t/T_1} \tag{3-3}$$

$$M_x=M_x(0)e^{-t/T_2} \tag{3-4}$$

Here $M_x(0)$ and $M_z(0)$ denote the respective values at time zero ($t=0$). The change of the magnetization in the direction of H_0, according to eq. (3-3), is achieved by the energy exchange between the spin system and the lattice. Hence, this process is called "spin–lattice relaxation" or "longitudinal relaxation," noting its direction. On the other hand, the relaxation in the direction perpendicular to H_0, according to eq. (3-4), is achieved by spin exchange within the spin system. This process is called "spin–spin relaxation" or "transverse relaxation," noting its direction.

In addition to the above-mentioned spin–lattice and spin–spin relaxations, there are many kinds of magnetic relaxation phenomena, such as the longitudinal relaxation in the rotating frame ($T_{1\rho}$) and nuclear Overhauser enhancement (NOE). All of these relaxations are caused by perturbation of the magnetic field in the vicinity of the spins. Any mechanism that initiates perturbation of a magnetic field, including a frequency component that corresponds to the energy splits in the spin system, can cause these relaxations. Examples are time-fluctuating dipole–dipole interactions, interaction of the spin with the electric quadrupole, interaction with the anisotropy of chemical shift, scalar coupling interaction, etc. Structural and conformational information generally can be obtained by examining such relaxation phenomena.

However, the relaxation of magnetization for macromolecules including only nuclei with spin 1/2, such as ^{13}C and ^{1}H, stems mostly from the time-fluctuating dipole–dipole interaction. Particularly in the ^{13}C NMR of macromolecules with zero-spin nuclei, except ^{13}C and ^{1}H, the dipole–dipole interaction between ^{13}C and the surrounding ^{1}H, is the primary mechanism for relaxation; this is because the interaction between the ^{13}C nuclei themselves is negligibly small owing to their very low natural abundance (1.1%). Furthermore, as is shown by eq. (3-7), the dipole–dipole interaction rapidly diminishes with increasing internuclear distance. Then, the relaxation can be thought to occur mainly via dipole–dipole interaction between chemically bonded ^{13}C and ^{1}H. Because the time fluctuation of the interaction takes place as a result of the time-fluctuation of the internuclear vector, examination of relaxation phenomena affords insight into molecular chain dynamics. Particularly with ^{13}C NMR instrumentation available at present, the chemical shifts corresponding to different carbons in molecular structure can be distinguished clearly for most macromolecular substances. Therefore, the mode of motion of the internuclear vector concerning individual carbons can be elucidated, and detailed information regarding the conformation and dynamics of the entire molecular chain can be determined.

The principle of magnetic relaxation caused by dipole–dipole interaction is well established and widely used for analyzing experimental relaxation data. However, there are very few books that describe the basic principles in detail. Hence, this chapter first introduces the basic theory of relaxation caused by dipole–dipole interaction.

Theory of Magnetic Relaxation by Dipole–Dipole Interaction[1–5]

Eigenstates of a Two-Spin System

Consider a system of two spins **I** and **S** in order to understand the mechanism of magnetic relaxation by the dipole–dipole interaction. For simplicity we deal with the case in which both spins are 1/2, ie, the maximum eigenvalues of their one-dimensional components, such as I_z, S_z or I_x, S_x are 1/2, where the subscripts indicate their respective directions. The reader may consider that **I** and **S** correspond to ^{1}H and ^{13}C, respectively. The Hamiltonian of this system under a static field H_0 in relation to the dipole–dipole relaxation can be expressed as the sum of the Zeeman energy term $\mathscr{H}_0$, the interaction of the spin system with H_0, and the dipole–dipole interaction term $\mathscr{H}_d$.

$$\mathscr{H} = \mathscr{H}_0 + \mathscr{H}_d \tag{3-5}$$

Because the interaction of a magnetic moment $\boldsymbol{\mu}$ with a magnetic field $\mathbf{H}$ is $-\boldsymbol{\mu}\cdot\mathbf{H}$ and the magnetic moments of the two spins are $\hbar\gamma_I\mathbf{I}$ and $\hbar\gamma_S\mathbf{S}$, $\mathscr{H}_0$ is given as:

$$\mathscr{H}_0 = -\hbar\gamma_I I_z H_0 - \hbar\gamma_S S_z H_0 \tag{3-6}$$

where γ_I and γ_S are the magnetogyric ratios of the respective spins.

According to classical electromagnetics, the field produced by a dipole moment $\boldsymbol{\mu}$ at a position $\mathbf{r}$ sufficiently removed from the dipole is $-(\boldsymbol{\mu}/r^3) + 3(\boldsymbol{\mu}\cdot\mathbf{r})\mathbf{r}/r^5$ [6]. Hence, the dipole–dipole interaction between **I** and **S** can be expressed as:

$$\mathscr{H}_d = \frac{\gamma_I\gamma_S\hbar^2}{r^3}\left[\mathbf{I}\cdot\mathbf{S} - \frac{3}{r^2}(\mathbf{I}\cdot\mathbf{r})(\mathbf{S}\cdot\mathbf{r})\right] \tag{3-7}$$

Here, $\mathbf{r}$ is the vector of length r connecting **I** and **S**. For ordinary NMR spectroscopy, the amplitude of $\mathscr{H}_0$ is much larger than that of $\mathscr{H}_d$, and $\mathscr{H}_d$ fluctuates with time according to the time-fluctuation of the vector $\mathbf{r}$ due to molecular motion. It then may be assumed that the unperturbed states of the system are determined by the Zeeman energy term $\mathscr{H}_0$, and $\mathscr{H}_d$ acts as a perturbation to produce the transition between these unperturbed states.

There are four unperturbed states determined by $\mathscr{H}_0$. These states are expressed by *cket* vectors $|++)$, $|+-)$, $|-+)$, and $|--)$. Here, the plus and minus signs indicate the states that give $\pm 1/2$, respectively, as the eigenvalues of z components of the spin **I** or **S**. The sign is written in relation to the spin **I** and **S** in order from left to right in the *cket* vectors. Namely, they are defined as:

$$I_z|+\beta) = \tfrac{1}{2}|+\beta),\ I_z|-\beta) = -\tfrac{1}{2}|-\beta)$$
$$S_z|\alpha+) = \tfrac{1}{2}|\alpha+),\ S_z|\alpha-) = -\tfrac{1}{2}|\alpha-) \tag{3-8}$$

where α denotes the sign + or − in relation to the state of **I** spin and β in relation to that of **S** spin.

These states are not only the eigenstates of $\mathscr{H}_0$ but also the eigenstates of the z components of the spins **I** and **S**, as revealed from eq. (3-6). They comprise, of course, an orthogonal system. Furthermore, they are assumed to be normalized as:

$$(\alpha'\beta'|\alpha\beta) = \delta_{\alpha'\beta',\alpha\beta} \tag{3-9}$$

Here, *bra* $(\alpha'\beta'|$, of course, is the conjugated vector to $|\alpha'\beta')$, and $(\alpha'\beta'|\alpha\beta)$ expresses the integrated result of the product of $(\alpha'\beta'|$ and $|\alpha\beta)$ in the entire space concerned; δ denotes Kronecker's delta ($\delta_{l,m}=1$ for $l=m$, $\delta_{l,m}=0$ for $l\neq m$).

In other words, the system consists of a four-dimensional vector space on the basis of the four *cket* vectors $|\alpha\beta)$. Provided $|a)$ and $|b)$ are *cket* vectors in this space, their inner product is defined as $(a|b)$ and the length (or magnitude) of $|a)$ is defined as the square root of $(a|a)$. Any state concerned can be treated as a *cket* vector in this space. For example, the eigenstates of the spin components perpendicular to H_0, such as I_x or S_x, can be expressed as vectors in this space. In regard to such transverse eigenstates that give $\pm 1/2$ as the eigenvalues of I_x and S_x, we can consider four states in a similar way to the longitudinal eigenstates $|\alpha\beta)$ cited above. Let us designate these transverse eigenstates by *cket* vectors $|_x\alpha\beta)$; then $|_x\alpha\beta)$ are defined as

$$I_x\,|_x+\beta)=\tfrac{1}{2}\,|_x+\beta),\ I_x\,|_x-\beta)=-\tfrac{1}{2}\,|_x-\beta)$$

$$S_x\,|_x\alpha+)=\tfrac{1}{2}\,|_x\alpha+),\ S_x\,|_x\alpha-)=-\tfrac{1}{2}\,|_x\alpha-) \qquad (3\text{-}10)$$

These *cket* vectors $|_x\alpha\beta)$ can be expressed as *cket* vectors in the above-mentioned vector space on the basis of $|\alpha\beta)$ as follows.*

$$\begin{aligned}|_x++)&=\tfrac{1}{2}\,[|++)+|+-)+|-+)+|--)]\\ |_x+-)&=\tfrac{1}{2}\,[|++)-|+-)+|-+)-|--)]\\ |_x-+)&=\tfrac{1}{2}\,[|++)+|+-)-|-+)-|--)]\\ |_x--)&=\tfrac{1}{2}\,[|++)-|+-)-|-+)+|--)]\end{aligned} \qquad (3\text{-}11)$$

These transverse eigenstates of the spin system, of course, are not the eigenstates of the energy but they are necessary when considering the transient transverse magnetization of the system.

Mathematical Formulation of Magnetic Relaxation of a Two-Spin System in Terms of Transition Probabilities among Energy Eigenstates

In order to obtain mathematical formulas for the longitudinal and transverse relaxations of the magnetization, the transition probabilities among the longitudinal and the transverse eigenstates discussed in the previous section are defined as w_1, w_1', w_0, w_2 and u_1, u_1', u_0, u_2 as shown in the Figure 3-1.

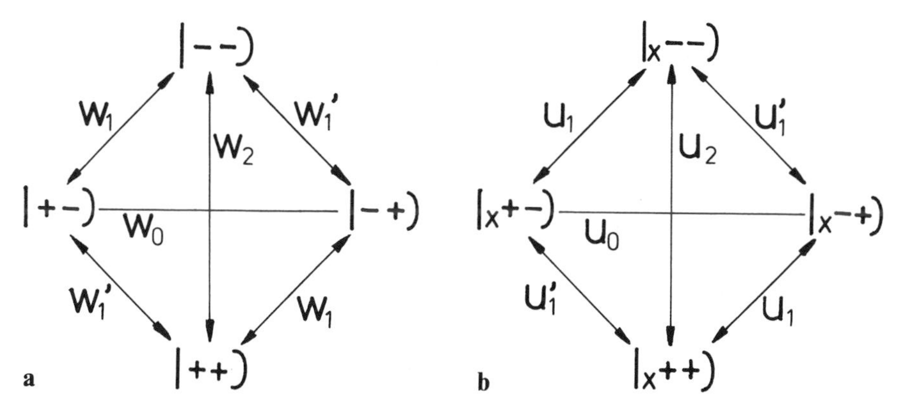

Figure 3-1. Longitudinal and transverse eigenstates and definition of transition probabilities among the states. (a) Longitudinal eigenstates, (b) transverse eigenstates.

*The reader can confirm that $|_x\alpha\beta)$ are the eigenstates of I_x and S_x as expressed by Eq. (3-11), replacing I_x and S_x as $I_x=(I_++I_-)/2$ and $S_x=(S_++S_-)/2$ and using the relations: $I_\pm|\mp\beta)=|\pm\beta)$, $I_\pm|\pm\beta)=0$, $S_\pm|\alpha\mp)=|\alpha\pm)$, $S_\pm|\alpha\pm)=0$. Here $I_\pm$ and $S_\pm$ are the so-called raising and lowering operators; defined as $I_\pm=I_x\pm iI_y$, $S_\pm=S_x\pm iS_y$.

Consider, first, the time dependency of the occupation probabilities or the populations of the spins at the energy eigenstates. Because the occupation probability $N_{\alpha\beta}$ at the state $|\alpha\beta)$ decreases with the transition to other states and increases with the transition from other states:

$$\frac{dN_{++}}{dt} = -(w_1 + w_1' + w_2)N_{++} + w_1' N_{+-} + w_1 N_{-+} + w_2 N_{--} + \text{constant} \tag{3-12a}$$

$$\frac{dN_{+-}}{dt} = w_1' N_{++} - (w_0 + w_1 + w_1')N_{+-} + w_0 N_{-+} + w_1 N_{--} + \text{constant} \tag{3-12b}$$

$$\frac{dN_{-+}}{dt} = w_1 N_{++} + w_0 N_{+-} - (w_0 + w_1 + w_1')N_{-+} + w_1' N_{--} + \text{constant} \tag{3-12c}$$

$$\frac{dN_{--}}{dt} = w_2 N_{++} + w_1 N_{+-} + w_1' N_{-+} - (w_1 + w_1' + w_2)N_{--} + \text{constant} \tag{3-12d}$$

Here, the constant terms are introduced by considering the energy exchange between the spin system and the lattice. Unless the spin system contacts the lattice, all occupation probabilities become identical and all magnetization disappears in the equilibrium state at infinite time. Because the expected values $\langle I_z \rangle$ and $\langle S_z \rangle$ of I_z and S_z are thought to be proportional to the excess occupation probabilities of the eigenstates that give positive eigenvalues over that of the eigenstates that give negative eigenvalues:

$$K\langle I_z \rangle = (N_{++} + N_{+-}) - (N_{-+} + N_{--}) \tag{3-13a}$$

$$K\langle S_z \rangle = (N_{++} + N_{-+}) - (N_{+-} + N_{--}) \tag{3-13b}$$

Here, K is a common constant for both equations.

Differentiating:

$$K\frac{d\langle I_z \rangle}{dt} = \frac{dN_{++}}{dt} + \frac{dN_{+-}}{dt} - \frac{dN_{-+}}{dt} - \frac{dN_{--}}{dt} \tag{3-14a}$$

$$K\frac{d\langle S_z \rangle}{dt} = \frac{dN_{++}}{dt} + \frac{dN_{-+}}{dt} - \frac{dN_{+-}}{dt} - \frac{dN_{--}}{dt} \tag{3-14b}$$

Substitution of eqs. (3–12) yields

$$K\frac{d\langle I_z \rangle}{dt} = -2(w_1 + w_2)(N_{++} - N_{--}) - 2(w_1 + w_0)(N_{+-} - N_{-+}) + \text{constant} \tag{3-15a}$$

$$K\frac{d\langle S_z \rangle}{dt} = -2(w_1' + w_2)(N_{++} - N_{--}) + 2(w_1' + w_0)(N_{+-} - N_{-+}) + \text{constant} \tag{3-15b}$$

Because the relations are derived from eqs. (3-13):

$$2(N_{++} - N_{--}) = K(\langle I_z \rangle + \langle S_z \rangle) \tag{3-16a}$$

$$2(N_{+-} - N_{-+}) = K(\langle I_z \rangle - \langle S_z \rangle) \tag{3-16b}$$

this gives

$$\frac{d\langle I_z \rangle}{dt} = -(w_0 + 2w_1 + w_2)\langle I_z \rangle - (w_2 - w_0)\langle S_z \rangle + \text{constant} \tag{3-17a}$$

$$\frac{d\langle S_z \rangle}{dt} = -(w_0 + 2w_1' + w_2)\langle S_z \rangle - (w_2 - w_0)\langle I_z \rangle + \text{constant} \tag{3-17b}$$

Let the equilibrium values of $\langle I_z \rangle$ and $\langle S_z \rangle$ be $\langle I_z \rangle_0$ and $\langle S_z \rangle_0$ when $d\langle I_z \rangle/dt = 0$ and $d\langle S_z \rangle/dt = 0$; then:

$$\frac{d\langle I_z \rangle}{dt} = -(w_0 + 2w_1 + w_2)(\langle I_z \rangle - \langle I_z \rangle_0) - (w_2 - w_0)(\langle S_z \rangle - \langle S_z \rangle_0) \tag{3-18a}$$

$$\frac{d\langle S_z \rangle}{dt} = -(w_0 + 2w_1' + w_2)(\langle S_z \rangle - \langle S_z \rangle_0) - (w_2 - w_0)(\langle I_z \rangle - \langle I_z \rangle_0) \tag{3-18b}$$

Replacing the sum of the transition probabilities as:

$$\rho = w_0 + 2w_1 + w_2 \tag{3-19a}$$

$$\rho' = w_0 + 2w_1' + w_2 \tag{3-19b}$$

$$\sigma = w_2 - w_0 \tag{3-19c}$$

eqs. (3-18) can be rewritten:

$$\frac{d\langle I_z \rangle}{dt} = -\rho(\langle I_z \rangle - \langle I_z \rangle_0) - \sigma(\langle S_z \rangle - \langle S_z \rangle_0) \tag{3-20a}$$

$$\frac{d\langle S_z \rangle}{dt} = -\rho'(\langle S_z \rangle - \langle S_z \rangle_0) - \sigma(\langle I_z \rangle - \langle I_z \rangle_0) \tag{3-20b}$$

Moreover, in relation to the time dependence of the expected values of the transverse components of spins **I** and **S**, similar treatments can be made. Noting in these cases that $\langle I_x \rangle$ and $\langle S_x \rangle = 0$ in the equilibrium states when $d\langle I_x \rangle/dt = 0$ and $d\langle S_x \rangle/dt = 0$, then:

$$\frac{d\langle I_x \rangle}{dt} = -(u_0 + 2u_1 + u_2)\langle I_x \rangle - (u_2 - u_0)\langle S_x \rangle \tag{3-21a}$$

$$\frac{d\langle S_x \rangle}{dt} = -(u_0 + 2u_1' + u_2)\langle S_x \rangle - (u_2 - u_0)\langle I_x \rangle \tag{3-21b}$$

Provided the transition probabilities among the eigenstates are time-independent constants, the time dependence of the expected values of the spin components in different directions can be formulated by solving the conjugated differential eqs. (3-20) and (3-21).

The consideration of the transition probabilities is important to understanding the magnetic relaxation. The explicit form of the probabilities will be given in the next section. Here we solve eqs. (3-20) and (3-21) for some cases, assuming the constancy of the transition probabilities.

(i) If the expected value of one of the paired spins is either zero or stays unchanged, the differential eqs. (3-20) can be solved easily. For example, in eq. (3-20b), if $\langle I_z \rangle = \langle I_z \rangle_0$, then:

$$\frac{d\langle S_z \rangle}{dt} = -\rho'(\langle S_z \rangle - \langle S_z \rangle_0) \tag{3-22}$$

Therefore:

$$\langle S_z \rangle - \langle S_z \rangle_0 = \text{constant} \times e^{-\rho' t} \tag{3-23}$$

The condition assumed here, ie, $\langle I_z \rangle = \langle I_z \rangle_0$, is not unrealistic. Consider the ^{13}C NMR spectroscopy of macromolecules containing ^{13}C and ^{1}H as nonzero spin nuclei. The magnetogyric ratio of ^{13}C is much smaller than that of ^{1}H (γ_H is about four times γ_C). Hence, the resonance electromagnetic wave can be applied selectively at the ^{13}C spin without affecting the ^{1}H spin, holding the equilibrium magnetization of the latter constant throughout the experiment. This condition therefore corresponds to a real experimental condition when ^{13}C longitudinal magnetic relaxation is observed, after an 180° pulse for example, without applying the resonance electromagnetic wave on ^{1}H. (Such spectroscopies without ^{1}H decoupling may be unusual, because the signal-to-noise ratio is pronouncedly reduced due to the lack of nuclear Overhauser enhancement. Nevertheless, ^{1}H decoupling can be omitted principally.)

Because experimentally observable magnetization corresponds to the expected value of the spin multiplying with $\gamma\hbar$, from eq. (3-23):

$${}^{c}M_z(t) - {}^{c}M_{z^0} = ({}^{c}M_z(0) - {}^{c}M_{z^0})e^{-\rho' t} \tag{3-24}$$

or, because ${}^{c}M_z(0) = -{}^{c}M_{z^0}$ for observing the relaxation after a 180° pulse, as is usually the case:

$${}^{c}M_z(t) - {}^{c}M_{z^0} = -2{}^{c}M_{z^0}e^{-\rho' t} \tag{3-25}$$

Here, ${}^{c}M_z$ and ${}^{c}M_{z^0}$ are ^{13}C magnetization in the z direction and its equilibrium value, respectively.

In another case, when the ^{13}C magnetic relaxation is observed under ^{1}H scalar decoupled conditions throughout the NMR experiment, ^{1}H magnetization is thought to be zero because of the rapid direction change. This real situation may correspond to the condition $\langle I_z \rangle = 0$ in eq. (3-20b). Then:

$$\frac{d\langle S_z \rangle}{dt} = -\rho'(\langle S_z \rangle - \langle S_z \rangle_0) + \sigma\langle I_z \rangle_0 \tag{3-26}$$

Integration with the note that ${}^{c}M_z = \gamma_c\hbar\langle S_z \rangle$, ${}^{H}M_{z^0} = \gamma_H\hbar\langle I_z \rangle_0$ yields:

$${}^{c}M_z(t) - \left[{}^{c}M_{z^0} + \frac{\sigma}{\rho'}\frac{\gamma_c}{\gamma_H}{}^{H}M_{z^0}\right] = \left\{{}^{c}M_z(0) - \left[{}^{c}M_z(0) + \frac{\sigma}{\rho'}\frac{\gamma_c}{\gamma_H}{}^{H}M_{z^0}\right]\right\}e^{-\rho' t} \tag{3-27}$$

Here the superscript H on the left indicates respective quantities for ^{1}H. Because $^{H}M_{z^0} = (\gamma_H^2/\gamma_c^2)^{c}M_{z^0}$ from eq. (3-1) (Curie's law):

$$^{c}M_z(t) - \left(1 + \frac{\sigma\gamma_H}{\rho'\gamma_c}\right)^{c}M_{z^0} = \left[^{c}M_z(0) - \left(1 + \frac{\sigma\gamma_H}{\rho'\gamma_c}\right)^{c}M_{z^0}\right]e^{-\rho' t} \quad (3\text{-}28)$$

(ii) In the other extreme case, if molecular motion is very much enhanced, simple relation between the parameters appearing in the conjugated eq. (3-20), $\rho \simeq \rho' \simeq 2\sigma$, is fulfilled. (This relation is obtained from eqs. (3-19) and (3-59) ~ (3-62) because (3-71) reduces to $J_m(\omega_{nn}) = 2\tau_c|\mathrm{Y}_m|^2$ for $\tau_c^2\omega^2 \ll 1$.)

Because $\rho' = \rho$, subtraction of eq. (3-20b) from eq. (3-20a) yields:

$$\frac{d(\langle I_z\rangle - \langle S_z\rangle)}{dt} = -(\rho - \sigma)(\langle I_z\rangle - \langle S_z\rangle) + (\rho - \sigma)(\langle I_z\rangle_0 - \langle S_z\rangle_0 \quad (3\text{-}29)$$

Defining a new valuable Ω, and its equilibrium value Ω_0 as $\Omega = \langle I_z\rangle - \langle S_z\rangle$, $\Omega_0 = \langle I_z\rangle_0 - \langle S_z\rangle_0$ for simplicity, then:

$$\frac{d\Omega}{dt} = -(\rho - \sigma)\Omega + (\rho - \sigma)\Omega_0 \quad (3\text{-}30)$$

Integration yields:

$$\Omega - \Omega_0 = (\text{constant}) \times e^{-(\rho - \sigma)t} \quad (3\text{-}31)$$

If a terminal condition: $\langle S_z\rangle = -\langle S_z\rangle_0$, $\langle I_z\rangle = \langle I_z\rangle_0$ at $t = 0$ (observation of the longitudinal relaxation of **S** spin by a 180°–τ–90° pulse) is introduced, then:

$$\Omega - \Omega_0 = 2\langle S_z\rangle_0 e^{-(\rho - \sigma)t} \quad (3\text{-}32)$$

Namely:

$$\langle I_z\rangle - \langle I_z\rangle_0 = \langle S_z\rangle - \langle S_z\rangle_0 + 2\langle S_z\rangle_0 e^{-(\rho - \sigma)t} \quad (3\text{-}33)$$

Substitution in eq. (3-20b) yields:

$$\frac{d\langle S_z\rangle}{dt} = -\rho(\langle S_z\rangle - \langle S_z\rangle_0) - \sigma[\langle S_z\rangle - \langle S_z\rangle_0 + 2\langle S_z\rangle_0 e^{-(\rho - \sigma)t}] \quad (3\text{-}34)$$

Integration yields:

$$\begin{aligned}\langle S_z\rangle - \langle S_z\rangle_0 &= -\langle S_z\rangle_0 e^{-(\rho - \sigma)t} - \langle S_z\rangle_0 e^{-(\rho + \sigma)t} \\ &= -\langle S_z\rangle_0 e^{-\rho t}(e^{\sigma t} + e^{-\sigma t})\end{aligned} \quad (3\text{-}35)$$

Furthermore, because $\sigma = \rho/2$, then:[3]

$$\begin{aligned}\langle S_z\rangle - \langle S_z\rangle_0 &= -\langle S_z\rangle_0 e^{-\rho t}(e^{\rho t/2} + e^{-\rho t/2}) \\ &= -2\langle S_z\rangle_0 e^{-\rho t}\cosh(\rho t/2)\end{aligned} \quad (3\text{-}36)$$

Here, $\cosh(\rho t/2) = (1/2)(e^{\rho t/2} + e^{-\rho t/2})$

In the region of shorter times, eq. (3-36) can be approximated to:

$$\langle S_z\rangle - \langle S_z\rangle_0 = -2\langle S_z\rangle_0 e^{-\rho t} \quad (3\text{-}37)$$

Therefore, when the longitudinal relaxation of ^{13}C spin is examined for substances in solution, eq. (3-25) is expected to be valid in the incipient state of the relaxation independent of the change of ^{1}H magnetization.

In either case, eqs. (3-24) and (3-28) are in the same form as eq. (3-3), which is experimentally established and discussed in the Introduction, with:

$$1/T_1 = \rho' = w_0 + 2w_1' + w_2 \tag{3-38}$$

The above discussion clearly indicates that the reciprocal of the experimentally obtainable spin–lattice relaxation time T_1 is equivalent to the sum of the transition probabilities among the eigenstates of the spin system. For the longitudinal magnetic relaxation of carbon bonded chemically with plural hydrogens, assuming the additivity of the relaxation mechanism between each pair of ^{13}C and ^{1}H, then:

$$1/nT_1 = \rho' = w_0 + 2w_1' + w_2 \tag{3-39}$$

Here, n is the number of hydrogens bonded chemically to the carbon concerned. Because (as is discussed in the following section) the transition probabilities are related intimately to molecular motion, examination of T_1 relaxation gives insight into the molecular motion of substances.

In the meantime, eq. (3-28) has important implications. Letting $t=\infty$:

$$^{c}M_z(\infty) = \left(1 + \frac{\sigma\gamma_H}{\rho'\gamma_c}\right) {}^{c}M_{z^0} \tag{3-40}$$

This shows that ^{13}C equilibrium longitudinal mangetization becomes greater by a factor of $1 + (\sigma\gamma_H/\rho'\gamma_c)$ than the equilibrium value when no decoupling irradiation is applied on ^{1}H. In other words, the ^{13}C equilibrium longitudinal magnetization is enhanced by applying the resonance electromagnetic wave on ^{1}H. This phenomenon is called nuclear Overhauser enhancement. Writing eq. (3-28) in the form:

$$^{c}M_z(t) - (1+\eta)^{c}M_{z^0} = [^{c}M_z(0) - (1+\eta)^{c}M_{z^0}]e^{-t/T_1} \tag{3-41}$$

the term $(1+\eta)$ is usually defined as nuclear Overhauser enhancement (NOE).

$$\mathrm{NOE} = 1 + \eta \tag{3-42}$$

with

$$\eta = \frac{\sigma\gamma_H}{\rho'\gamma_c} \tag{3-43}$$

Likewise, η is defined as the nuclear Overhauser enhancement factor. Equation (3-66) gives NOE in terms of spectral densities by calculating σ and ρ' and in more realistic form by calculating the spectral densities, assuming a particular mode of motion of C–H vector, such as in eq. (3-75). Note here that the NOEs are not affected either by the number of hydrogens bonded to the carbon concerned or by the internuclear distance.

On the other hand, in the conjugated differential eq. (3-21), in relation to the transverse magnetizations it may be assumed that the expected value of one of the paired spins is always zero. Letting $\langle I_x \rangle = 0$ in eq. (3-21b), therefore:

$$\langle S_x \rangle = \text{constant} \times e^{-(u_0 + 2u_1' + u_2)t} \tag{3-44}$$

Assuming that **S** spin corresponds to ^{13}C and **I** to ^{1}H spin, the equation for relaxation of ^{13}C transverse magnetization is obtained:

$$^{c}M_x = {}^{c}M_x(0)e^{-(u_0 + 2u_1' + u_2)t} \tag{3-45}$$

Comparing with eq. (3-4) the content of T_2 can be obtained:

$$1/T_2 = u_0 + 2u_1' + u_2 \tag{3-46}$$

For a carbon bonded with plural hydrogens,

$$1/nT_2 = u_0 + 2u_1' + u_2 \tag{3-47}$$

The relaxation parameters T_1, T_2 and NOE therefore can be correlated with the transition probabilities among the energy eigenstates by eqs. (3-39), (3-42), and (3-47). Hence, if the transition probabilities can be correlated to molecular motion, the measurement of such relaxation parameters eventually gives insight into molecular motion.

To this end, note that the basic equations for the longitudinal and the transverse ^{13}C magnetizations:

$$M_z(t) - M_z(\infty) = [M_z(0) - M_z(\infty)]e^{-t/T_1} \tag{3-48}$$

$$M_x(t) = M_x(0)e^{-t/T_2} \tag{3-49}$$

are theoretically valid only for a few definite cases, as discussed above. Equation (3-48) is generally believed to be valid for ^{13}C longitudinal magnetization of substances in solution. Nevertheless, the condition $\rho = \rho' = 2\sigma$, which guarantees the equation, in the incipient stage of the relaxation, is not always fulfilled for macromolecular substances, even in solution. Therefore, it is preferable to choose adequate conditions under which ^{1}H longitudinal magnetization is zero or stays constant during the relaxation by applying proper ^{1}H decoupling irradiations. Otherwise, one must take into account the change of ^{1}H magnetization. Also eq. (3-49) must be used carefully in evaluating the transverse relaxation time T_2, remarking the change of ^{1}H magnetization during the relaxation.

Transition Probabilities and Spectral Densities

In order to understand the content of the relaxation parameters T_1, T_2, and NOE, the transition probabilities among the eigenstates of the spin system must be considered. The transition probability per unit time, $w_{nn'}$ from a state n to another state n' by a perturbation Hamiltonian $\mathcal{H}_d$ can be expressed to the first approximation as:[7]

$$w_{n,n'} = \frac{1}{\hbar^2} \int_{-\infty}^{\infty} \overline{(n'|\mathscr{H}_d^*(t+\tau)|n)(n|\mathscr{H}_d(t)|n')}\, e^{-i\omega_{nn'}\tau}\, d\tau \qquad (3\text{-}50)$$

Here, $\omega_{nn'}$ is the energy difference between the two states expressed in frequency scale. Namely, let the energies of the state n and n' be $E_{n'}$ and E_n; then:

$$\omega_{nn'} = (E_{n'} - E_n)/\hbar \qquad (3\text{-}51)$$

The upper bar in eq. (3-50) indicates the average of the quantity in an ensemble of spin systems. Using this relation, the transition probabilities w_1, w_1', w_2, w_0, u_1, u_1', u_2 and u_0 that appeared in the previous section can be calculated to define the relaxation parameters. The interaction Hamiltonian $\mathscr{H}_d$ between spin **I** and **S** can be described as follows,[1] defining the interspin vector **r** as shown in the Figure 3-2:

$$\mathscr{H}_d = \gamma_I \gamma_S \hbar^2 \sum_m Y_m A_m \qquad (3\text{-}52)$$

with $m = 0, \pm 1, \pm 2$. Here, Y_m are the functions of vector **r** and are called orientation functions because they depend only upon the orientation of **r**.

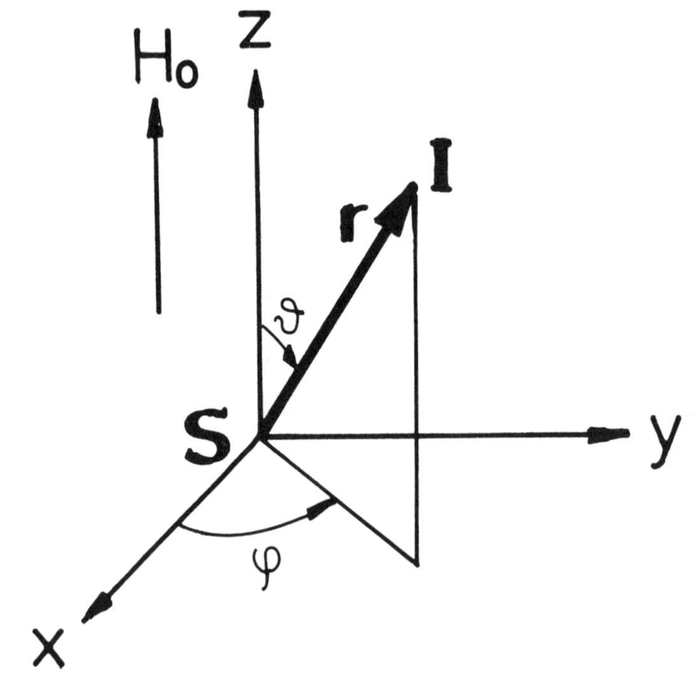

Figure 3-2. Rectangular and polar coordinates to define interspin vector **r**.

$$Y_0 = r^{-3}(1 - 3\cos^2\theta)$$
$$Y_{\pm 1} = r^{-3}\sin\theta\cos\theta\, e^{\pm i\phi}$$
$$Y_{\pm 2} = r^{-3}\sin^2\theta\, e^{\pm 2i\phi} \qquad (3\text{-}53)$$

In eq. (3-52), A_m are operator terms independent of $\mathbf{r}$ and, hence, independent of time, defined as:

$$A_0 = I_z S_z - \frac{1}{4}(I_+ S_- + I_- S_+)$$

$$A_{\pm 1} = -\frac{3}{2}(I_{\mp} S_z + I_z S_{\mp})$$

$$A_{\pm 2} = -\frac{3}{4} I_{\mp} S_{\mp} \qquad (3\text{-}54)$$

In these equations, the double signs are in order and the raising and lowering operators $I_\pm$ and $S_\pm$ are defined as $I_\pm = I_x \pm iI_y$, $S_\pm = S_x \pm iS_y$ (see the footnote of eq. (3-11)). If these relations are inserted in eq. (3-50) the cross-terms in relation to m disappear, because $A_{\pm m}$ are Hermite conjugates of $A_{\mp m}$, and $Y_{\pm m}$ are complex conjugates of $Y_{\mp m}$. Then:

$$w_{n,n'} = \gamma_I^2 \gamma_S^2 \hbar^2 \sum_m |(n'|A_m|n)|^2 \int_{-\infty}^{\infty} \overline{Y_m^*(t+\tau) Y_m(t)} \exp(-i\omega_{nn'}\tau)\, d\tau \qquad (3\text{-}55)$$

Here, Y_m^* is the complex conjugate of Y_m. The correlation functions of the orientation functions $\overline{Y_m^*(t+\tau)Y_m(t)}$ depend only on the time difference τ, insofar as the ensemble average in a stational state is concerned. Defining spectral densities $J_m(\omega_{nn'})$ to be proportional to the Fourier transforms of the correlation functions as:

$$J_m(\omega_{nn'}) = \int_{-\infty}^{\infty} \overline{Y_m^*(t+\tau) Y_m(t)} \exp(-i\omega_{nn'}\tau)\, d\tau \qquad (3\text{-}56)$$

Note that the spectral densities are sometimes defined as:

$$J'_m(\omega_{nn'}) = \int_{-\infty}^{\infty} \overline{Y_m^*(t+\tau) Y_m(t)} \exp(-i\omega_{nn'}\tau)\, d\tau / \overline{|Y_m(t)|^2}$$

because the value at $\tau = 0$ of the correlation function $\overline{Y_m^*(t+\tau)Y_m(t)}$ is independent of t insofar as it is concerned with the ensemble average in a stational state, and is given usually by eq. (3-70) as will be shown below. In this case, note that the coefficients in the following equations including the spectral densities must be changed appropriately.

Equation (3-55) is rewritten in a compact form:

$$w_{n,n'} = \gamma_I^2 \gamma_S^2 \hbar^2 \sum_m |(n'|A_m|n)|^2 J_m(\omega_{nn'}) \qquad (3\text{-}57)$$

Here, $J_m(\omega_{nn'}) = J_{-m}(\omega_{nn'})$.

The time-independent operator terms $|(n'|A_m|n)|^2$ can be calculated easily for the case that $|n)$, $|n')$ are the energy eigenstates defined by eq. (3-8), ie, $|++)$, $|+-)$, $|-+)$, and $|--)$, noting that: $I_{\pm}|\mp\beta)=|\pm\beta)$, $I_{\pm}|\pm\beta)=0$, $S_{\pm}|\alpha\mp)=|\alpha\pm)$, $S_{\pm}|\alpha\pm)=0$. Therefore:

$$|(\alpha\beta|A_0|\alpha\beta|^2=\frac{1}{16},\quad |(\mp\pm|A_0|\pm\mp)|^2=\frac{1}{16},$$

$$|(\mp\beta|A_{\pm 1}|\pm\beta)^2=\frac{9}{16},\quad |(\alpha\mp|A_{\pm 1}|\alpha\pm)|^2=\frac{9}{16},$$

$$|(\mp\mp|A_{\pm 2}|\pm\pm)|^2=\frac{9}{16},\quad \text{and}\quad |(\alpha'\beta'|A_m|\alpha\beta)|^2=0 \tag{3-58}$$

for all pair of $\alpha\beta$ and $\alpha'\beta'$ unless described above. Accordingly, the transition probabilities among the energy eigenstates shown in Figure 3-1 are calculated as:

$$\begin{aligned} w_0 &= w_{|+-),|-+)} = w_{|-+),|+-)} \\ &= \gamma_I^2\gamma_S^2\hbar^2\sum_m|(-+|A_m|+-)|^2 J_m(\omega_I-\omega_S) \\ &= \gamma_I^2\gamma_S^2\hbar^2|(-+|A_0|+-)|^2 J_0(\omega_I-\omega_S) \\ &= \frac{1}{16}\gamma_I^2\gamma_S^2\hbar^2 J_0(\omega_I-\omega_S) \end{aligned} \tag{3-59}$$

$$\begin{aligned} w_1 &= w_{|+-),|--)} = w_{|--),|+-)} = w_{|++),|-+)} = w_{|-+),|++)} \\ &= \gamma_I^2\gamma_S^2\hbar^2\sum_m|(--|A_m|+-)|^2 J_m(\omega_I) \\ &= \frac{9}{16}\gamma_I^2\gamma_S^2\hbar^2 J_1(\omega_I) \end{aligned} \tag{3-60}$$

$$w_1' = \frac{9}{16}\gamma_I^2\gamma_S^2\hbar^2 J_1(\omega_S) \tag{3-61}$$

$$\begin{aligned} w_2 &= w_{|++),|--)} = w_{|--),|++)} \\ &= \frac{9}{16}\gamma_I^2\gamma_S^2\hbar^2 J_2(\omega_I+\omega_S) \end{aligned} \tag{3-62}$$

where ω_I and ω_S are the Larmor frequencies of spin **I** and **S**, respectively. Substitution in eqs. (3-19) yields:

$$\rho' = \frac{1}{16}\gamma_I^2\gamma_S^2\hbar^2[J_0(\omega_I-\omega_S)+18J_1(\omega_S)+9J_2(\omega_I+\omega_S)] \tag{3-63}$$

$$\sigma = \frac{1}{16}\gamma_I^2 I\gamma_S^2\hbar^2[9J_2(\omega_I+\omega_S)-J_0(\omega_I-\omega_S)] \tag{3-64}$$

Substitution of ρ', σ in eqs. (3-39) and (3-41), replacing γ_S and γ_I by γ_C and γ_H, respectively, yields the formulas for T_1 and NOE of ^{13}C:

$$\frac{1}{nT_1} = \frac{1}{16}\gamma_H^2\gamma_C^2\hbar^2[J_0(\omega_H - \omega_C) + 18J_1(\omega_C) + 9J_2(\omega_H + \omega_C)] \quad (3\text{-}65)^*$$

$$\text{NOE} = 1 + \frac{9J_2(\omega_H + \omega_C) - J_0(\omega_H - \omega_C)}{J_0(\omega_H - \omega_C) + 18J_1(\omega_C) + 9J_2(\omega_H + \omega_C)} \cdot \frac{\gamma_H}{\gamma_C} \quad (3\text{-}66)^*$$

The transition probabilities among the transverse eigenstates u_0, u_1, u_1', and u_2, defined in Figure 3-2, also can be calculated using an equation similar to eq. (3-50),[3] because the transverse eigenstates are not energy eigenstates but are expressed by the superposed sum of the energy eigenstates as eqs. (3-11). Therefore, we have the formula for the spin–spin relaxation time T_2 of ^{13}C in terms of spectral densities:

$$\frac{1}{nT_2} = \frac{1}{32}\gamma_H^2\gamma_C^2\hbar^2[4J_0(0) + J_0(\omega_H - \omega_C) + 18J_1(\omega_C)$$
$$+ 36J_1(\omega_H) + 9J_2(\omega_H + \omega_C)] \quad (3\text{-}67)^*$$

The relaxation parameters T_1, T_2, and NOE therefore are described in terms of the spectral densities, ie, in terms of Fourier transforms of the correlation functions of the orientation functions. Because the orientation functions are, of course, a function of internuclear vector **r** as defined by eqs. (3-53), examination of these relaxation parameters gives information about the time dependence of the internuclear vector and, hence, molecular motion in principle.

Nuclear Spin Relaxation When the Internuclear Vectors Undergo Spherically Random Rotation (Single Correlation-Time Theory)

Consider the case where the internuclear vector **r** changes direction at random spherically while holding its length r. The orientation functions defined by eq. (3-53) will fluctuate through random time dependence of θ and ϕ. In this case, the correlation functions will be given by a correlation time constant τ_c

*If the spectra densities are defined as noted in relation to eq. (3-56), eqs. (3-65)–(3-67) must be rewritten as:

$$\frac{1}{nT_1} = \frac{\gamma_H^2\gamma_C^2\hbar^2}{20r^6}[J_0'(\omega_H - \omega_C) + 3J_1'(\omega_C) + 6J_2'(\omega_H + \omega_C)] \quad (3\text{–}65)'$$

$$\frac{1}{nT_2} = \frac{\gamma_H^2\gamma_C^2\hbar^2}{40r^6}[4J_0'(0) + J_0'(\omega_H - \omega_C) + 3J_1'(\omega_C)$$
$$+ 6J_1(\omega_H) + 6J_2(\omega_H + \omega_C)] \quad (3\text{–}66)'$$

$$\text{NOE} = 1 + \frac{6J_2'(\omega_H + \omega_C) - J_0'(\omega_H - \omega_C)}{J_0'(\omega_H - \omega_C) + 3J_1'(\omega_C) + 6J_2'(\omega_H + \omega_C)} \cdot \frac{\gamma_H}{\gamma_C} \quad (3\text{–}67)'$$

common for all orientation functions as:[1,2]

$$\overline{Y^*_{\rm m}(t+\tau)Y_{\rm m}(t)} = \overline{Y^*_{\rm m}(t)Y_{\rm m}(t)}\, e^{-|\tau|/\tau_c}$$
$$= \overline{|Y_{\rm m}(t)|^2}\, e^{-|\tau|/\tau_c} \qquad (3\text{-}68)$$

As pointed out before in relation to eq. (3-55) as far as the ensemble-average of the spin systems in a stational state is concerned, the correlation functions depend only upon the time difference τ, independent of t. Their initial values $\overline{|Y_{\rm m}(t)|^2}$ are also independent of t and are described as:

$$\overline{|Y_{\rm m}|^2} = \int_0^{2\pi} d\phi \int_0^{\pi} d\theta \sin\theta\, Y^*_{\rm m}(r,\theta,\phi)Y_{\rm m}(r,\theta,\phi) \Big/ \int_0^{2\pi} d\theta \int_0^{\pi} \sin\theta\, d\theta \qquad (3\text{-}69)$$

Using eq. (3-53) with constant r, the $\overline{|Y_{\rm m}|^2}$ are calculated as:

$$\overline{|Y_0|^2} = (4/5)r^{-6}$$
$$\overline{|Y_{\pm 1}|^2} = (2/15)r^{-6}$$
$$\overline{|Y_{\pm 2}|^2} = (8/15)r^{-6} \qquad (3\text{-}70)$$

Therefore, the spectral densities defined by eq. (3-56) are obtained in the form:

$$J_{\rm m}(\omega_{\rm nn'}) = \int_{-\infty}^{\infty} \overline{|Y_{\rm m}|^2}\, e^{-|\tau|/\tau_c} e^{-i\omega_{\rm nn'}\tau}\, d\tau$$
$$= \overline{|Y_{\rm m}|^2} \frac{2\tau_c}{1+\tau_c^2\omega_{\rm nn'}^2} \qquad (3\text{-}71)$$

$$J_0(\omega_{\rm nn'}) = \frac{8}{5} r^{-6} \frac{\tau_c}{1+\tau_c^2\omega_{\rm nn'}^2} \qquad (3\text{-}72a)$$

$$J_1(\omega_{\rm nn'}) = \frac{4}{15} {\rm r}^{-6} \frac{\tau_c}{1+\tau_C^2\omega_{\rm nn'}^2} \qquad (3\text{-}72b)$$

$$J_2(\omega_{\rm nn'}) = \frac{16}{15} {\rm r}^{-6} \frac{\tau_c}{1+\tau_c^2\omega_{\rm nn'}^2} \qquad (3\text{-}72c)$$

From eqs. (3-65)–(3-67) the formulas for T_1, T_2, and NOE of ^{13}C are obtained:

$$1/nT_1 = \frac{\gamma_H^2\gamma_C^2\hbar^2}{10r^6}\left[\frac{\tau_c}{1+\tau_c^2(\omega_H-\omega_C)^2} + \frac{3\tau_c}{1+\tau_c^2\omega_C^2} + \frac{6\tau_c}{1+\tau_c^2(\omega_H+\omega_C)^2}\right] \qquad (3\text{-}73)$$

$$1/nT_2 = \frac{\gamma_H^2\gamma_C^2\hbar^2}{20r^6}\left[4\tau_c + \frac{\tau_c}{1+\tau_c^2(\omega_H-\omega_C)^2} + \frac{3\tau_c}{1+\tau_c^2\omega_C^2} + \frac{6\tau_c}{1+\tau_c^2\omega_H^2} + \frac{6\tau_c}{1+\tau_c^2(\omega_H+\omega_C)^2}\right] \qquad (3\text{-}74)$$

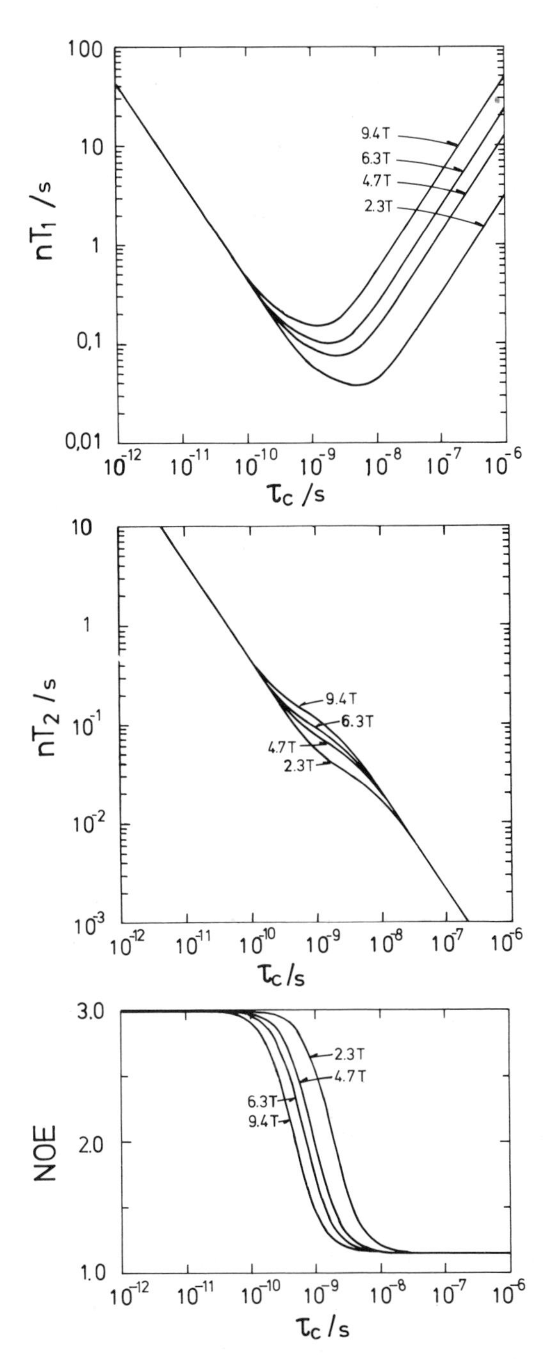

Figure 3-3. Carbon-13 T_1, T_2, and NOE versus correlation time according to the single correlation-time theory at different field strengths for $r = 1.08$ Å; n denotes the number of protons bonded to ^{13}C. The field strengths are indicated in Tesla to each curve. (a) nT_1 vs τ_c, (b) nT_2 vs τ_c, (c) NOE vs τ_c.

$$\mathrm{NOE} = 1 + \frac{\left[\dfrac{6}{1+\tau_c^2(\omega_H+\omega_C)^2}\right] - \left[\dfrac{1}{1+\tau_c^2(\omega_H-\omega_C)^2}\right]}{\left[\dfrac{1}{1+\tau_c^2(\omega_H-\omega_C)^2}\right] + \left[\dfrac{3}{1+\tau_c^2\omega_C^2}\right] + \left[\dfrac{6}{1+\tau_c^2(\omega_H+\omega_C)^2}\right]} \cdot \frac{\gamma_H}{\gamma_C} \tag{3-75}$$

In the extremely narrowing limit $\tau_c^2(\omega_H+\omega_C)^2 \ll 1$, these formulas simplify to:

$$1/nT_1 \cong \frac{\gamma_H^2\gamma_C^2\hbar^2\tau_c}{r^6} \tag{3-76}$$

$$1/nT_2 \cong \frac{\gamma_H^2\gamma_C^2\hbar^2\tau_c}{r^6} \tag{3-77}$$

$$\mathrm{NOE} \cong 1 + \frac{\gamma_H}{2\gamma_C} = 2.99 \tag{3-78}$$

The relationships between the magnetic relaxation times, the correlation time of molecular motion, and the Larmor frequencies of ^{13}C and ^{1}H described by eqs. (3-73)–(3-75) are shown in Figure 3-3. The values of nT_1, nT_2, and NOE are plotted against the correlation time τ_c at four static magnetic fields of 2.3, 4.7, 6.3, and 9.4 T, ie, ^{1}H Larmor frequencies of 100, 200, 270, and 400 MH$_z$ for r=1.08Å). It can be seen that the frequency dependence of T_1 at longer correlation times and the consistency of T_1 and T_2 at shorter correlation times independent of the frequency. These relations are useful for describing ^{13}C magnetic relaxation phenomena of monomeric substances to the first approximation when their molecular motion is pronounced, as is the case in solution or in the molten state. However, ^{13}C magnetic relaxation phenomena observed for macromolecular substances seldom follow these relations, as to be noted in the experimental sections, even when the predominant mechanism of relaxation is dipole–dipole interaction. In the case of polymers, the motion of internuclear vectors between carbon and proton are thought to be very anisotropic, characteristic of long-chain molecules. Nevertheless, the relaxation phenomena of polymers are mostly considered using the equations.

Nuclear Spin Relaxation when an Ellipsoidal Molecule Undergoes Rotational Random Motion

As pointed out in the previous section, the theory based on the spherical random rotation of internuclear vectors does not describe exactly the nuclear spin relaxation phenomena not only for macromolecules but also for monomeric substances when the internuclear vectors undergo an anisotropic motion. In such cases, however, if it is assumed that the internuclear vectors undergo a particular anisotropic motion, the relaxation phenomena can be described formally. There are many models as anisotropic motion of internuclear vectors.

The first of these is a simple case in which an ellipsoidal molecule is assumed to undergo rotational motion in which the internuclear vector **r** under consideration is embedded.[8] Consider a rectangular coordinate system S', of which x, y, z axes coincide with three axes of the ellipsoid, respectively. Let l, m, n be the direction cosines of the internuclear vector **r** in this frame S'. Assume that the frame S' undergoes random diffusional rotation such that three independent random rotations occur around the x, y, z axes with rotational diffusion coefficients R_x, R_y, R_z, respectively. At an arbitrary time t, the frame S' may be thought to orient randomly to the laboratory frame for an ensemble of isotropically oriented molecules. Then, the ensemble-averaged correlation functions of the orientation functions $\overline{Y_m^*(t+\tau)Y_m(t)}$ can be formulated as a function of the time difference τ, independent of arbitrary time t, in terms of the rotational diffusional coefficients:[8]

$$\overline{Y_m^*(t+\tau)Y_m(t)} = \frac{1}{2}K_m[C_+e^{-|\tau|/\tau_+} + C_-e^{-|\tau|/\tau_-} + C_1e^{-|\tau|/\tau_1} + C_2e^{-|\tau|/\tau_2} + C_3e^{-|\tau|/\tau_3}] \tag{3-79}$$

where K_m are the average of the correlation function at an arbitrary time t, described by eq. (3-70):

$$K_m = \overline{|Y_m|^2} \tag{3-80}$$

and

$$1/\tau_1 = 4R_x + R_y + R_z \tag{3-81a}$$

$$1/\tau_2 = R_x + 4R_y + R_z \tag{3-81b}$$

$$1/\tau_3 = R_x + R_y + 4R_z \tag{3-81c}$$

$$1/\tau_\pm = 2(R_x + R_y + R_z) \pm 2[(R_x + R_y + R_z)^2 - 3(R_xR_y + R_yR_z + R_zR_x)]^{1/2} \tag{3-82}$$

$$C_1 = 6m^2n^2 \tag{3-83a}$$

$$C_2 = 6l^2n^2 \tag{3-83b}$$

$$C_3 = 6l^2m^2 \tag{3-83c}$$

$$C_\pm = \frac{1}{2}[3(l^4 + m^4 + n^4) - 1] \mp \frac{1}{6}[\delta_1(3l^4 + 6m^2n^2 - 1) + \delta_2(3m^4 + 6l^2n^2 - 1) + \delta_3(3n^4 + 6l^2m^2 - 1)] \tag{3-84}$$

with

$$\delta_1 = \frac{3R_x - (R_x + R_y + R_z)}{[(R_x + R_y + R_z)^2 - 3(R_xR_y + R_yR_z + R_zR_x)]^{1/2}} \tag{3-85a}$$

$$\delta_2 = \frac{3R_y - (R_x + R_y + R_z)}{[(R_x + R_y + R_z)^2 - 3(R_xR_y + R_yR_z + R_zR_x)]^{1/2}} \tag{3-85b}$$

$$\delta_3 = \frac{3R_z - (R_x + R_y + R_z)}{[(R_x + R_y + R_z)^2 - 3(R_xR_y + R_yR_z + R_zR_x)]^{1/2}} \tag{3-85c}$$

In eq. (3-79), if the direction cosines of the internuclear vector **r** in the frame S'

are assumed to be independent of τ, ie, if the vector $\mathbf{r}$ is fixed in the molecular ellipsoid, substitution of eq. (3-79) into eq. (3-56) yields the spectral densities $J_m(\omega)$:

$$J_m(\omega) = K_m\left[C_+\frac{\tau_+}{1+\omega^2\tau_+^2} + C_-\frac{\tau_-}{1+\omega^2\tau_-^2} + C_1\frac{\tau_1}{1+\omega^2\tau_1^2} + C_2\frac{\tau_2}{1+\omega^2\tau_2^2} + C_3\frac{\tau_3}{1+\omega^2\tau_3^2}\right] \quad (3\text{-}86)$$

Hence, substitution of the spectral densities into eqs. (3-65)–(3-67) yields the relaxation times T_1, T_2 and the NOE in terms of three rotational diffusion coefficients R_x, R_y, R_z with l, m and n, ie, the orientation of vector $\mathbf{r}$ in the ellipsoid. Consider the extreme case wherein rotational motion is much faster than the measurement frequency, such as $\omega^2\tau_i^2 \ll 1$; then T_1 and T_2 coincide:

$$1/T_1 = 1/T_2 = \frac{1}{2}\gamma_H^2\gamma_C^2\hbar^2 r^{-6}\sum_n [{}^nC_+\tau_+ + {}^nC_-\tau_- + {}^nC_1\tau_1 + {}^nC_2\tau_2 + {}^nC_3\tau_3] \quad (3\text{-}87)$$

Here, the summation should be carried out in relation to different internuclear vectors if plural hydrogens are bonded to the carbon concerned. This situation is the same as in the single correlation-time theory presented in the previous section.

Equation (3-86) describes the spectral densities when the ellipsoid, in which the internuclear vector is embedded at a fixed position, involves three independent diffusional rotations with different diffusional coefficients.

In the case of an axially symmetrical ellipsoid, two of the three diffusional coefficients should be taken as equivalent such that $R_{\|} = R_x$, $R_{\perp} = R_y = R_z$. In this case the frame S' can be chosen as $n=0$, $m^2 = 1 - l^2$. Then, the parameters in eq. (3-86) reduce to

$$\tau_1 = (4R_{\|} + 2R_{\perp})^{-1}, \quad \tau_2 = \tau_3 = (R_{\|} + 5R_{\perp})^{-1}$$

$$\tau_+ = (4R_{\|} + 2R_{\perp})^{-1}, \quad \tau_- = (6R_{\perp})^{-1}$$

$$C_+ = \frac{1}{2}(1 - l^2)^2, \quad C_- = \frac{1}{2}(1 - 3l^2)^2 \quad (3\text{-}88)$$

Then eq. (3–87) can be simplified:

$$J_m(\omega) = 2K_m\left[A\frac{\tau_A}{1+\omega^2\tau_A^2} + B\frac{\tau_B}{1+\omega^2\tau_B^2} + C\frac{\tau_C}{1+\omega^2\tau_C^2}\right] \quad (3\text{-}89)$$

where

$$A = \frac{1}{4}(3l^2 - 1)^2 \quad (3\text{-}90a)$$

$$B = 31^2(1 - l^2) \quad (3\text{-}90b)$$

$$C = \frac{3}{4}(l^2 - 1)^2 \quad (3\text{-}90c)$$

with $A + B + C = 1$:

$$1/\tau_A = 6R_\perp \tag{3-91a}$$

$$1/\tau_B = R_\parallel + 5R_\perp \tag{3-91b}$$

$$1/\tau_C = 4R_\parallel + 2R_\perp \tag{3-91c}$$

Note here that if $R_\parallel = R_\perp$, eq. (3-89) becomes identical to eq. (3-71), replacing τ_c by $\tau_A = \tau_B = \tau_C$. Equation (3-89) holds regardless of whether $R_\parallel > R_\perp$ or $R_\parallel < R_\perp$ and the spectral densities, and hence the relaxation times as well, depend only on the direction cosine l of the internuclear vector to the x axis of the ellipsoid ($R_\parallel$ is the rotational diffusion constant around the x axis in S'), independent of the direction cosines m and n.

The formulation for the axially symmetrical ellipsoid is of particular importance because of the simplicity and applicability to some kinds of proteins with such an ellipsoidal molecular shape.

Furthermore, in the extreme limit: $\omega^2\tau_i^2 \ll 1$, eq. (3-89) with eqs. (3-65) and (3-67) yields

$$\frac{1}{nT_1} = \frac{1}{nT_2} = \frac{\gamma_H^2\gamma_C^2\hbar^2}{r^6}\left[\frac{A}{6R_\perp} + \frac{B}{R_\parallel + 5R_\perp} + \frac{C}{4R_\parallel + 2R_\perp}\right] \tag{3-92}$$

whereas eq. (3-78) also holds for NOE in this case.

For convenience, one can introduce the ratio:

$$\sigma = R_\parallel / R_\perp \tag{3-93}$$

Then, eq. (3-92) can be rewritten:

$$\frac{1}{nT_1} = \frac{1}{nT_2} = \frac{\gamma_H^2\gamma_C^2\hbar^2}{r^6} R_\perp^{-1}\left[\frac{A}{6} + \frac{B}{(\sigma + 5)} + \frac{C}{(4\sigma + 2)}\right] \tag{3-94}$$

Using this equation, the anisotropic effect of the rotational diffusion motion on the relaxation parameters T_1 and T_2 can be evaluated. Equation (3-94) is identical to eqs. (3-76) and (3-77) in the case of the single correlation-time theory, but systematic decrease in $1/nT_1$ and $1/nT_2$, ie, increase in T_1 and T_2, is anticipated with increasing σ, depending upon A, B, and C, ie, depending on the direction cosine l (see Figure 1 in reference 8).

Because the anisotropy of the rotational motion σ can be correlated with the anisotropy of the ellipsoid, ie, the ratio of the long axis to the short axis, the effect of the anisotropy of molecules on the relaxation times will be evaluated correctly if the molecular shape can be known by other appropriate measurements.[8]

The theories introduced here on the basis of the diffusional rotation of ellipsoid embedding the internuclear vector improve the single correlation-time theory and describe relaxation phenomena in many cases for monomeric substances, even when the latter fails. However, for relaxation phenomena of macromolecules, these theories are sometimes still insufficient. The internuclear vector in macromolecules is thought to involve much more highly anisotropic motions.

Hitherto, this discussion has assumed that the internuclear vector is fixed in the ellipsoid, irrespective of whether the ellipsoid undergoes spherical or nonspherical diffusional rotations. These models are sometimes denoted as "rigid rotor models". As a more realistic model for macromolecules, however, internal motion of the internuclear vector within the ellipsoid also should be assumed. For example, for the axially symmetric ellipsoid model, if it is assumed that the internuclear vector rotates stochastically about the rotational symmetrical axis of the elipsoid with a fixed angle, the spectral densities are given by:

$$J_m(\omega) = 2K_m\left[A\frac{\tau_{A'}}{1+\omega^2\tau_{A'}^2} + B\frac{\tau_{B'}}{1+\omega^2\tau_{B'}^2} + C\frac{\tau_{C'}}{1+\omega^2\tau_{C'}^2}\right] \quad (3\text{-}95)$$

where:

$$1/\tau_{A'} = 1/\tau_A = 6R_\perp \quad (3\text{-}96a)$$

$$1/\tau_{B'} = 1/\tau_B + R_{x'} = R_\parallel + 5R_\perp + R_{x'} \quad (3\text{-}96b)$$

$$1/\tau_{C'} = 1/\tau_C + 4R_{x'} = 4R_\parallel + 2R_\perp + 4R_{x'} \quad (3\text{-}96c)$$

Here, τ_A, τ_B, τ_C and A, B, C denote the same quantities as in eq. (3-89). If the angle of the internuclear vector to the rotational symmetrical axis of the ellipsoid (x axis) is designated as Δ, $l = \cos^2\Delta$. Then:

$$A = \frac{1}{4}(3\cos^2\Delta - 1)^2 \quad (3\text{-}97a)$$

$$B = \frac{3}{4}\sin^2\Delta \quad (3\text{-}97b)$$

$$C = \frac{3}{4}\sin^4\Delta \quad (3\text{-}97c)$$

where $R_{x'}$ is the internal rotational diffusion coefficient of $\mathbf{r}$ about the x axis in the frame of the ellipsoid S'.

Furthermore, if the ellipsoid undergoes isotropic random rotational motion, such as $R_\parallel = R_\perp$, then $\tau_{A'}$, $\tau_{B'}$, and $\tau_{C'}$ in eq. (3-95) are given by:

$$\tau_{A'}^{-1} = \tau_I^{-1} \quad (3\text{-}98a)$$

$$\tau_{B'}^{-1} = \tau_I^{-1} + \tau_{R'}^{-1} \quad (3\text{-}98b)$$

$$\tau_{C'}^{-1} = \tau_I^{-1} + 4\tau_{R'}^{-1} \quad (3\text{-}98c)$$

where τ_I characterizes the random spherical rotational motion of the ellipsoid ($\tau_I^{-1} = 6R_\perp = 6R_\parallel$), and $\tau_{R'}$ the stochastic inner rotational motion of vector $\mathbf{r}$ around an axis of the ellipsoid ($\tau_{R'} = R_{x'}^{-1}$). Here it is assumed that when the vector $\mathbf{r}$ is located as an arbitrary time t at an azimuthal angle ϕ_0 to the axis of the ellipsoid, the probability of finding it at time $t+\tau$ at $\phi_0+\Delta\phi$ is expressed as:

$$p(\Delta\phi, \tau) = \frac{1}{4}(\pi R_{x'}\tau)^{-1/2} \exp(-\Delta\phi^2/4R_{x'}\tau) \quad (3\text{-}99a)$$

or

$$p(\Delta\phi, \tau) = \frac{1}{4}(\pi\tau/\tau_{R'})^{-1/2} \exp(-\Delta\phi^2\tau_{R'}/4\tau) \quad (3\text{-}99b)$$

If, however, the vector **r** is assumed to undergo random jump rotation about the axis between three equilibrium positions at an average rate of $(3\tau_{R'})^{-1}$, $\tau_{C'}$ in eq. (3-95) becomes identical to $\tau_{B'}$ so that eqs. (3-98) should be replaced by:[9]

$$\tau_{A'}^{-1} = \tau_I^{-1} \quad (3\text{-}100a)$$

$$\tau_{B'}^{-1} = \tau_{C'}^{-1} = \tau_I^{-1} + \tau_{R'}^{-1} \quad (3\text{-}100b)$$

The above-described relaxation theory based on the rotational diffusion model of the ellipsoid may be effective, as long as the molecular shape of substances examined can be defined, irrespective of whether the molecule really exhibits ellipsoidal shape or not. Particularly, the ellipsoid model, including the internal rotation of the internuclear vector, is applicable to some macromolecules. In fact, many peptides or proteins are known that possess a definite molecular shape (eg, helical, spherical, or ellipsoidal) in solution or in their natural states. In such cases, the above-noted relaxation theory has been found to be very effective. Some experimental examples are presented in a later section, Carbon-13 Spin Relaxation of α-Carbons in Proteins.

Molecular Motion with Plural Correlation Times, Pertinent to Flexible Long-Chain Macromolecules

As long as the entire molecular shape is definable, as is the case for many peptides and proteins, the ellipsoid rotational model may be useful. However, flexible long-chain macromolecules seldom exhibit such definite molecular chain conformation, and a partinent relaxation model is needed for such cases. Consider a particular case for the ellipsoid rotational model, where the ellipsoid undergoes diffusional rotations isotropically (R_I) and the internuclear vector further undergoes diffusional rotation about an axis of the ellipsoid with an angle Δ and a rotational diffusion constant $R_{x'}$ (or correlation time $\tau_{R'} = 1/R_{x'}$). The spectral densities in this case are given by eq. (3-95) with eqs. (3-97) and (3-98), and the relaxation parameters T_1, T_2, and NOE can be formulated with use of eqs. (3-65)–(3-67).

This model corresponds to Woessner's 2τ rotation model[9] shown in Figure 3-4(a), where axis $\overline{os}'$ undergoes a spherical rotation and vector **r** further rotates stochastically about $\overline{\mathbf{os}}'$ axis with angle Δ. Here, it may be considered that the $\overline{\mathbf{os}}'$ axis determines the direction of a segment of a flexible chain molecule in random conformation. Hence, this model may be applicable also to flexible long-chain macromolecules.

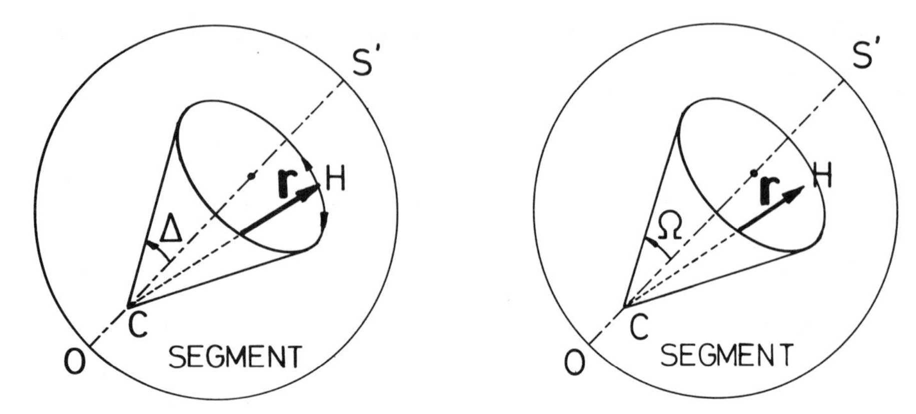

Figure 3-4. Schematic representation of 2-τ models. (a) Woessner's 2τ rotation model; (b) Howarth's 2τ librational rotation model. *O*, origen; other parameters defined in text.

In this model the stochastic diffusional rotation of **r** is assumed. However, if another mode of internal motion is assumed, the spectral densities are formulated differently. For example, the spectral densities are formulated by Howarth[10,11] for a case in which the vector **r** librates about the axis $\overline{\mathbf{os}}'$ within a cone defined by an angle Ω, as shown in Figure 3-4(b), as:

$$J_m(\omega) = 2K_m\left[A'\frac{\tau_I}{1+\omega^2\tau_I^2} + (1-A')\frac{\tau_H}{1+\omega^2\tau_H^2}\right] \tag{3-101}$$

with

$$A' = \frac{(\cos\Omega - \cos^3\Omega)^2}{4(1-\cos\Omega)^2} \tag{3-102}$$

$$\tau_H^{-1} = \tau_I^{-1} + \tau_L^{-1} \tag{3-103}$$

Here τ_I characterizes the spherical rotation of $\overline{os'}$ axis as in the case of Woessner's 2τ rotation model, and R_L the librational motion of vector **r** about $\overline{\mathbf{os}}'$. This model is designated here as Howarth's 2τ libration model. These 2τ rotation or libration models, including the internal motions of **r** with isotropic spherical segmental rotation of the molecule, formally describe the ^{13}C spin relaxation phenomena of long-chain macromolecules in some cases. It was found in the analysis of T_1 observed for different polyesters[12] and linear polyethylene[13,14] that these models accounted for the temperature dependence of T_1 for individual carbons. However, for different carbons, different values of the parameters (eg, τ_I, τ_R, or τ_L) and different modes of motion should be assumed. Individual carbons, of course, can exhibit different internal motions, but as long as they belong to a same long-chain molecule, the basic parameter τ_I, which characterizes the random segmental motion of the entire molecular chain, is considered to be the same for all individual carbons. Hence, these models are still unsatisfactory for explaining the motion of flexible long-chain macromolecules.

As a more realistic model for long-chain molecules, Howarth proposed a model connecting the 2τ rotation model with the 2τ libration model as shown in Figure 3-5.[11] For this model, spectral densities are proposed as a function of three correlation times and two vertical angles in the form:

$$J_m(\omega) = K_m\left[AA'\frac{2\tau_I}{1+\omega^2\tau_I^2} + A(1-A')\frac{2\tau_a}{1+\omega^2\tau_a^2} + BA'\frac{2\tau_b}{1+\omega^2\tau_b^2} + B(1-A')\frac{2\tau_d}{1+\omega^2\tau_d^2} + CA'\frac{2\tau_e}{1+\omega^2\tau_e^2} + C(1-A')\frac{2\tau_f}{1+\omega^2\tau_f^2}\right] \quad (3\text{-}104)$$

Here, A, B, C, and A' are the functions of the vertical angle Ω and Δ in eqs. (3-97) and (3-102) and:

$$\tau_a^{-1} = \tau_I^{-1} + \tau_L^{-1} \quad (3\text{-}105a)$$

$$\tau_b^{-1} = \tau_I^{-1} + \tau_{R'}^{-1} \quad (3\text{-}105b)$$

$$\tau_d^{-1} = \tau_I^{-1} + \tau_L^{-1} + \tau_{R'}^{-1} \quad (3\text{-}105c)$$

$$\tau_e^{-1} = \tau_I^{-1} + 4\tau_{R'}^{-1} \quad (3\text{-}105d)$$

$$\tau_f^{-1} = \tau_I^{-1} + \tau_L^{-1} + 4\tau_{R'}^{-1} \quad (3\text{-}105e)$$

Here, τ_I is a correlation time characterizing the spherical diffusional rotation of axis $\overline{os'}$, τ_L the librational motion of the $\overline{o'C}$ axis about the $\overline{os'}$ axis within the cone with angle Ω, and $\tau_{R'}$ the stochastic rotation of the internuclear vector **r** about the $\overline{o'C}$ axis. This equation is not established theoretically but can be derived as follows:

Fourier transform of eq. (3-95) yields the correlation functions:

$$\overline{F_m^*(t+\tau)F_m(t)} = K_m[Ae^{-|\tau|/\tau_{A'}} + Be^{-|\tau|/\tau_{B'}} + Ce^{-|\tau|/\tau_{C'}}] \quad (3\text{-}106)$$

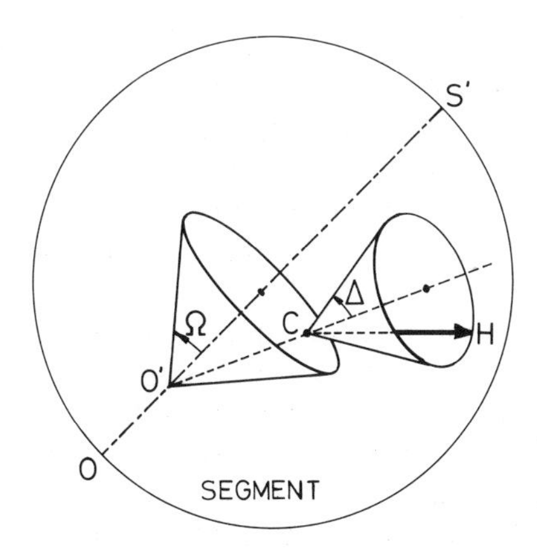

Figure 3-5. Schematic representation of Howarth's 3τ model.

Substitution of eq. (3-100) yields:

$$\overline{F^*_{\mathrm{m}}(t+\tau)F_{\mathrm{m}}(t)} = K_{\mathrm{m}}e^{-|\tau|/\tau_{\mathrm{I}}}[A + Be^{-|\tau|/\tau_{\mathrm{R'}}} + Ce^{-4|\tau|/\tau_{\mathrm{R'}}}] \tag{3-107}$$

Similarly, eq. (3-101) can be rewritten:

$$\overline{F^*_{\mathrm{m}}(t+\tau)F_{\mathrm{m}}(t)} = K_{\mathrm{m}}[A'e^{-|\tau|/\tau_{\mathrm{I}}} + (I - A')e^{-|\tau|(\tau_{\mathrm{I}}^{-1}+\tau_{\mathrm{L}}^{-1})}] \tag{3-108}$$

Substitution of the right-hand term of eq. (3-108) in the term, $K_{\mathrm{m}}\, e^{-|\tau|/\tau_{\mathrm{I}}}$, of eq. (3-107) yields:

$$\begin{aligned}\overline{F^*_{\mathrm{m}}(t+\tau)F(t)} = K_{\mathrm{m}}[&AA'e^{-|\tau|/\tau_{\mathrm{I}}} + A(1-A')e^{-|\tau|(\tau_{\mathrm{I}}^{-1}+\tau_{\mathrm{L}}^{-1})} \\ &+ BA'e^{-|\tau|(\tau_{\mathrm{I}}^{-1}+\tau_{\mathrm{R'}}^{-1})} + B(1-A')e^{-|\tau|(\tau_{\mathrm{I}}^{-1}+\tau_{\mathrm{L}}^{-1}+\tau_{\mathrm{R'}}^{-1})} \\ &+ CA'e^{-|\tau|(\tau_{\mathrm{I}}^{-1}+4\tau_{\mathrm{R'}}^{-1})} + C(1-A')e^{-|\tau|(\tau_{\mathrm{I}}^{-1}+\tau_{\mathrm{L}}^{-1}+4\tau_{\mathrm{R'}}^{-1})}]\end{aligned} \tag{3-109}$$

There is no theoretical base that the substitution of $K_{\mathrm{m}} \exp(-|\tau|/\tau_{\mathrm{I}})$ in eq. (3-107) by the right-hand term of eq. (3-108) yields the orientation functions and, then, the spectral densities for this 3τ rotation model. Hence, eq. (3-104) is not valid strictly. The spectral densities for this model have been strictly formulated by the authors.[15] It has been verified that the equation (3-104) is still valid if the coefficient A′ is somewhat corrected. It is noted that if $\tau_{\mathrm{I}} \gg \tau_{\mathrm{L}} \gg \tau_{\mathrm{R}}$, either eq. (3-104) or the correct equation reduces to in the form:

$$J_{\mathrm{m}}(\omega) = K_{\mathrm{m}}\left[A_1\frac{\tau_{\mathrm{I}}}{1+\omega^2\tau_{\mathrm{I}}^2} + B_2\frac{\tau_{\mathrm{L}}}{1+\omega^2\tau_{\mathrm{L}}^2} + A_3\frac{\tau_{\mathrm{R}}}{1+\omega^2\tau_{\mathrm{R}}^2} + A_4\frac{\tau_{\mathrm{R}}}{1+\omega^2(\tau_{\mathrm{R}}/2)^2}\right]$$

where A_1, A_2, A_3 and A_4 are a function of Δ and Ω. This suggests that the anisotropic 3τ rotation model discussed here is actually equivalent to a simple model involving three independent isotropic rotations with well-separated different correlation times.

Other Modes of Motion of Internuclear Vector

Hitherto, only a few models have been introduced for random motions of the internuclear vector to describe the spectral densities and, hence, the spin relaxation times. However, many other models have been proposed and discussed in relation to the actual molecular motion of substances. To review all of them is beyond the scope of this chapter; only brief mention will be made here. It is noted first that, in the ellipsoidal rotation model, the internal motion of the internuclear vector could be modeled differently: for example, eqs. (3-95) with (3-96) or (3-98) describe the spectral densities when the internuclear vector undergoes stochastic rotation about one axis of the ellipsoid in addition to the spheroidal or spherical diffusional rotation of the ellipsoid itself. In a similar way the spectral densities can be formulated if different types of internal motions are assumed, such as hindered rotation,[16] or random jump rotation with equivalent[8,17,18,19] or nonequivalent[20,21] probabilities between two or three sites about one axis of the ellipsoid.

The internal motion of the internuclear vector is not always restricted to such random rotations about one axis: an example has been shown in the previous section, where the internal motion is comprised of two independent librational and rotational motions. For models that include many multiple independent internal motions, the spectral densities may be formulated generally using Wigner's rotation matrix.[22] Such a complex model sometimes is needed to describe relaxation phenomena of carbons in side-chain residue of macromolecules.[23] In fact the ^{13}C-spin–lattice relaxation times of individual carbons of the rather long and flexible side chain of poly(N^5-(3-hydroxypropyl L-glutamine) are well analyzed by the model, assuming further random spherical rotation for the main chain motion with a distribution of correlation time, as will be reviewed briefly in the following experimental section, Motion of Side Chains in Polypeptides.[24]

Finally a model for molecular motion of macromolecules should be mentioned that can account for spin relaxation phenomena on the basis of "de Gennes' scaling concepts".[25,26] This model is based on the assumption that a given long-chain macromolecule is confined in a surrounding tube by entanglement with other chain molecules; rapid molecular motion is allowed within the tube and the tube configuration changes slowly. Assuming three independent stochastic molecular motions, ie, (a) defect reorientation with environmental fluctuation (rapid anisotropic segment reorientation), (b) longitudinal chain diffusion (de Gennes' reptation), and (c) configurational deformations of the surrounding tube, then the spectral densities for the same kind of spin system with spin 1/2 are derived and applied successfully to some ^1H spin relaxation phenomena of macromolecules in the melt.[27,28] Nevertheless, such an approach to understanding ^{13}C spin relaxation phenomena has seldom been made.[29]

Distribution of Correlation Time

As is well known, the concept of the distribution of relaxation time has been introduced occasionally in order to describe other relaxation phenomena, such as mechanical and dielectric relaxations observed for macromolecular substances. Also, for spin relaxation phenomena, particularly of carbons in the main chain of macromolecules, attempts have been made to describe them by use of single correlation-time theory, introducing a distribution in the correlation time.[30-34] In this case, the spectral densities $J_m(\omega)$ are described with a distribution function $G(\tau_c)$ of the correlation time instead of eq. (3-71) as:

$$J_{\mathrm{m}}(\omega) = 2\overline{|Y_{\mathrm{m}}|^2} \int_0^\infty \frac{\tau_c G(\tau_c)}{1+\tau_c^2\omega^2}\, d\tau_c \tag{3-110}$$

with $\int_0^\infty G(\tau_c)\, d\tau_c = 1.$

The distribution of correlation time was introduced in order to account for a deviation of the spin relaxation phenomena observed for viscous solutions of polymers or solid elastomers from that expected from the single correlation-time theory. For such viscous systems of macromolecules it is observed that ^{13}C T_1 are generally larger with very much lower T_2 than are expected from theory. Also, the NOEs barely approach 3 even at higher temperatures where molecular motion is enhanced.

Nevertheless, it was found by the authors that the spin relaxation phenomena observed for rubbery components in semicrystalline polymers, such as poly-(hexamethylene terephthalate) and linear polyethylene, could not be accounted for except by introducing a distribution in the correlation time.[12] For example, as shown in Figure 3-12, the minimum values of nT_1 at 60–100°C for methylene carbons of poly(hexamethylene terephthalate) are about 200 ms larger than expected from single correlation-time theory. In order to account for such larger nT_1, an extraordinarily wide distribution of correlation times should be introduced, regardless of the type of distribution.[35] However, such wide distribution was found to be inadequate because its introduction resulted in extreme flattening of the nT_1 versus temperature curve so that nT_1 becomes almost independent of average correlation time and, hence, independent of temperature.

What is the physical meaning of the distribution in the correlation time? It means that the correlation function of the orientation functions of the internuclear vector does not follow eq. (3-68) but follows:

$$\overline{Y_m^*(t+\tau)Y_m(t)} = \overline{|Y_m(t)|^2} \int_0^\infty G(\tau_c)e^{-|\tau|/\tau_c}\, d\tau_c \qquad (3\text{-}111)$$

There is no theoretical reason that the correlation function of the orientation functions of an ensemble of spin systems comprising a monophase should follow an exponential decay as described by eq. (3-68). Nevertheless, introduction of an extraordinarily wide distribution in the correlation time is not intuitively satisfying as long as we are concerned with a homogeneous phase even if a particular situation for long-chain macromolecules having a wide distribution of molecular weights is explicitly considered.

Otherwise, this problem can be assumed to be treated in terms of a complex phase composed of innumerable phases; the correlation function of the orientation functions of each phase follows eq. (3-68) with individual correlation times. In this case, different values must be assumed for ρ' defined by eqs. (3-19) for each phase. Thus, the right-hand term of eq. (3-25) becomes a summation over different values of ρ' and, hence, the longitudinal relaxation deviates continuously from the simple exponential decay. However, such a continuous departure from exponential decay is not observed in most experiments.

There may be some physical meaning for introducing a not so wide distribution in the correlation time even for an ensemble of spin pairs comprising a homogeneous phase, provided that the phase is so viscous that averaging of reorientation of the interspin vector is very slow. Nevertheless, a particular mode of molecular motion even for relaxation phenomena of main-chain carbons should be considered before a distribution in the correlation time is introduced.

Applications

General Aspect of ^{13}C Spin Relaxation and Interpretation by Single Correlation-Time Theory

The range of chemical shifts in ^{13}C NMR is generally wide enough so that individual carbons in ordinary macromolecules can be easily distinguished even under relatively low magnetic fields. Figure 3-6 shows the ^{1}H decoupled ^{13}C NMR spectrum of the ethylene–hexene-1 copolymer obtained by Hama et al[36] in tetrachloroethylene solution at 100°C and a static field of 2.3 T. Five types of backbone carbons (one CH and four CH_2 groups) as well as four types of side-chain carbons (one CH_3 and three CH_2 groups) can be identified readily.

Here, chemical shifts of individual carbons from TMS are indicated in brackets in each position of the chemical structure. Insofar as such chemical shifts are distinguishable, the spin relaxation times, such as T_1, of each carbon can be obtained by ordinary pulse sequences. Figure 3-7 shows T_1 of the same copolymer obtained by Axelson et al[37] in 40% trichlorobenzene solution at 118°C and 6.3 T.

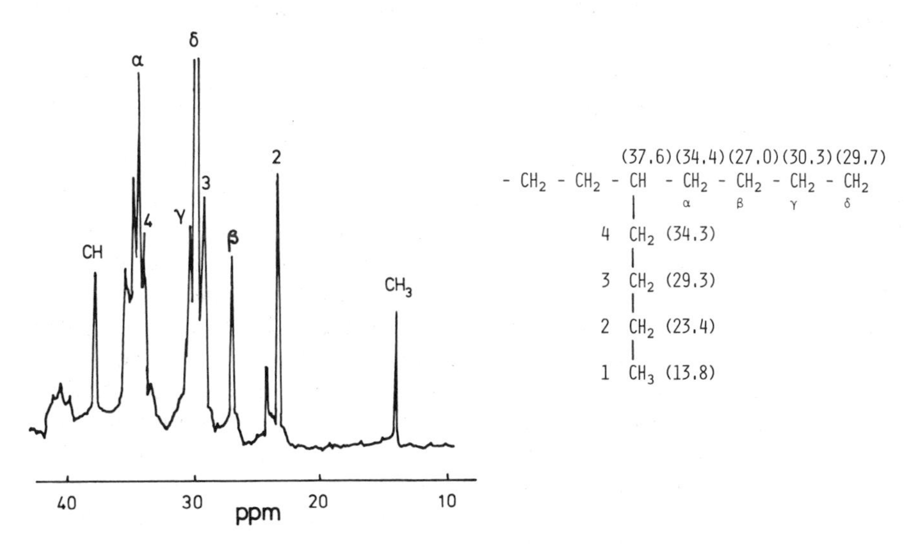

Figure 3-6. Carbon-13 spectrum of ethylene hexene-1 copolymer in 10% tetrachloroethylene solution at 100°C and 2.3 T. The chemical shift from tetramethylsilane (TMS) is indicated (in ppm) for each carbon. (From ref. 36 with permission).

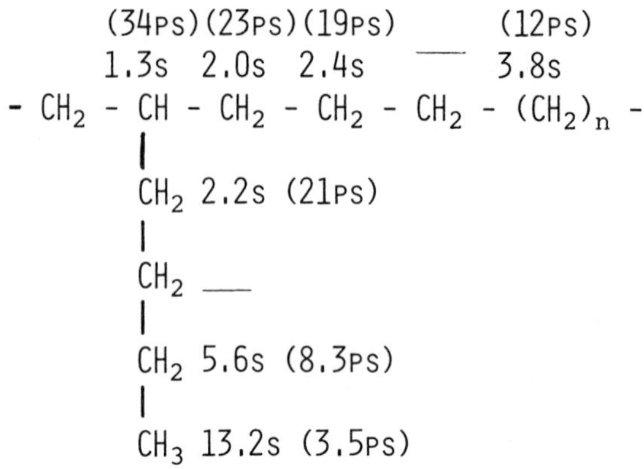

```
            (34ps)(23ps)(19ps)        (12ps)
             1.3s  2.0s  2.4s   ——     3.8s
  - CH2 - CH - CH2 - CH2 - CH2 - (CH2)n -
          |
          CH2 2.2s (21ps)
          |
          CH2 ——
          |
          CH2 5.6s (8.3ps)
          |
          CH3 13.2s (3.5ps)
```

Figure 3-7. The nT_1 of individual carbons of ethylene–hexene-1 copolymer in 40% trichlorobenzene solution. Data obtained at 118°C and 6.3 T. The value of τ_c calculated by eq. (3-76) are given (in ps). (From ref. 37, with permission.)

Here, nT_1 are indicated in each carbon position with the correlation times τ_c (in brackets), which are calculated according to eq. (3-73) or (3-76) with $r = 1.09A$ for CH_2 and CH_3 and $r = 1.08A$ for CH carbon. Note that the τ_c of backbone carbons are significantly larger than those of side-chain carbons and that the τ_c of both backbone and side-chain carbons become smaller as the distance from the branch site (CH) increases. The result reflects the fact that the molecular mobility of side-chain carbons is larger than that of the backbone carbons, and that both types of carbons, whether belonging to the backbone or to the side chain, involve different contributions to the segmental motion of the entire macromolecule because of differing degrees of conformational freedom. As can be seen by this example, the interpretation using the single correlation-time theory is not always useless if the extreme narrowing condition is fulfilled because of pronounced molecular motion as occurs in solution.

As another example suggesting the rationality of the single correlation-time theory, refer to T_1 data obtained by Inoue et al[38] for polyisobutylene samples possessing different molecular weights in solution at 40°C and 2.3 T. Table 3-1 summarizes the data of T_1 with nT_1 and correlation times τ_c calculated by eq. (3-76). The T_1 of each carbon are virtually the same for all molecular weight samples within experimental error except for the case of the smallest molecular weight sample. This result may imply that the T_1 observed here where the extreme narrowing condition is fulfilled represent only local molecular motion and that they are insensitive to the segmental motion of the entire molecular chain. It is also noted that nT_1 of CH_3 are sufficiently larger than those of CH_2 carbon. The mobility of each carbon can be evaluated roughly by examining the corresponding correlation time τ_c. The τ_c of CH_3 carbon are shorter than those of backbone CH_2 by a factor of 2. This result suggests that the molecular motion of the side group CH_3 is faster than that of the backbone CH_2 groups, presumably because of an additional stochastic rotation in the CH_3 group.

TABLE 3-1. PROTON-DECOUPLED ^{13}C SPIN–LATTICE RELAXATION AND CORRELATION TIME BASED ON SINGLE CORRELATION-TIME THEORY[a]

Molecular weight	T_1 (s)			nT_1 (s) (and τ_c (10^{-10} s))[b]	
	$-CH_2-$	$-C-$	$-CH_3$	$-CH_2-$	$-CH_3$
1,355	0.199	1.79	0.305	0.40 (1.2)	0.92 (0.51)
2,750	0.172	1.63	0.249	0.34 (1.4)	0.75 (0.62)
200,000	0.168	1.51	0.241	0.34 (1.4)	0.72 (0.65)
700,000	0.161	1.51	0.236	0.32 (1.5)	0.71 (0.66)

[a]Data in the table are taken from ref. 38 with permission.
[b]τ_c calculated by Eq. (3-76).
[c]Experimental error ±10%.

Information regarding the detailed molecular motion of macromolecules also can be obtained using nuclear Overhauser enhancement as discussed in the theoretical sections. Table 3-2 summarizes the data on NOE and T_1 obtained by Schaefer and Natusch[39] for different polymers. It can be seen that the NOEs are virtually the same for different types of carbon if compared within one kind of polymer, (eg, the data for polystyrene). Here, the NOEs are 1.8 for all sorts of carbons. As pointed out in the end of the Transition Probabilities and Spectral Densities section, the value of NOE should be independent of the number of protons coupled to the concerned carbon and the internuclear distances. For the NOE to be observable, only the presence of coupled protons is required. When considering whether or not an equivalent degree of molecular motion is involved, the NOEs should be equivalent regardless of whether the nuclei being observed are directly chemically bonded or not insofar as the internuclear C–H distance is held unchanged. The fact recognized here that the NOE of the nonprotonated carbon is equal to the NOEs of protonated carbons CH and CH_2 not only is in agreement with this theoretical expectation, but it also suggests that the ^{13}C spin relaxation concerned here arises primarily via the dipole–dipole interaction mechanism, as discussed by Kuhlmann et al.[40] If other relaxation mechanisms were operating in this case, the NOE of the nonprotonated carbon should be much smaller than those of CH and CH_2 carbons.

In Table 3-2, the correlation times calculated from both NOEs and T_1 are listed. Note that the correlation times from T_1 are appreciably smaller than those from NOEs. The NOEs hardly reach the maximum value 2.99 even when pronounced molecular motion occurs (as is suggested by the large values of T_1). Such a discordance is also widely recognized in the relation of T_1 to the spin–spin relaxation time T_2. If the extreme narrowing condition is fulfilled, T_1 and T_2 are expected to be identical by eqs. (3-76) and (3-77). Nevertheless, in most experiments involving macromolecules, the T_2 are found to be much smaller than the corresponding T_1 even when pronounced molecular motion is involved as, eg, in solution or in the molten state. These facts suggest inadequacy of interpretation of macromolecular relaxation phenomena by the single correlation-time theory.

TABLE 3-2. CARBON-13 NUCLEAR OVERHAUSER ENHANCEMENT AND T_1 OBTAINED AT 40°C AND 2.1 T WITH CORRELATION TIMES CALCULATED ON THE BASIS OF SINGLE CORRELATION-TIME THEORY BY EQS. (3-73) AND (3-75) FOR POLYMERS[a]

Polymer	Solvent	Carbon	NOE	T_1 (ms)	τ_c (from NOE) (10^{-9} s)	τ_c (from T_1) (10^{-9} s)
Poly(ethylene oxide)	Neat liquid	Methylene	2.7	265	1.0	0.088
Polyacrylonitrile	DMSO	Methine	2.2	115	1.7	0.38
		Methylene	2.2	60	1.7	0.39
Polystyrene (isotactic)	*o*-Dichlorobenzene	Quarternary	1.8	550	2.7	—
		Methine	1.8	65	2.7	0.68
		Methylene	1.8	32	2.7	0.73

[a]From ref. 39 with permission.

As cited above, the interpretation of T_1 or NOEs with the single correlation-time theory provides some information pertaining to molecular motion in relation to the chain conformation of macromolecules. It must be remembered, however, that the conclusion obtained via this theory is valid only to the first approximation. A higher order approximation is required in order to obtain detailed chain motion characteristics of macromolecules.

Carbon-13 Spin Relaxation of α Carbons in Proteins

Proton-decoupled ^{13}C spectra of proteins are generally well separated into groups of resonances arising, in order of increasing shielding, from carbonyl groups, unsaturated side chains, α carbons, and saturated side-chain residues even at a relatively low magnetic field, such as 1.41 T.[41] Among these groups of carbons, consider the spin relaxation of α carbons (backbone CH) and the protonated side-chain carbons (β carbons, γ carbons, etc.), which are known to arise via the ^{13}C–^{1}H dipole–dipole interaction mechanism. The dominance of the dipole–dipole interaction is reported as a relaxation mechanism for protonated carbons and some nonprotonated carbons as well in macromolecules;[42] the increasing importance of the effect of chemical shift anisotropy is reported for nonprotonated unsaturated carbons at higher magnetic fields.[43] The chemical shifts of protein α carbons are observable only as an envelope of those in a relatively narrow range of chemical shift. Nevertheless, the T_1 can be determined by the semilogarithmic plot of the integrated contribution of the α carbon envelope in the partially relaxed spectra against relaxed time τ in the usual pulse sequence (180°–τ–90°). Table 3-3 summarizes the data reported by Allerhand et al[44] for T_1 of horse myoglobin, human hemoglobin, hen egg-white lysozyme, and bovine serum albumin at the field strengths of 1.42 and 6.34 T. The correlation times, τ_R, are calculated according to eq. (3-73) with $\tau_R \equiv \tau_c$, $n=1$, $r=1.09A$, (of the two τ_R obtained from each T_1 the larger is employed). As pointed out in the section, Nuclear Spin Relaxation when an Ellipsoidal Molecule Undergoes Rotational Random Motion, in relation to eq. (3-89), the τ_R shown here characterize the diffusional rotation of an assumed spherical

TABLE 3-3. T_1 OF α CARBON ENVELOPES OF PROTEINS AND CORRELATION TIMES τ_R CALCULATED WITH THE USE OF ISOTROPIC RIGID ROTOR MODEL[a]

		T_1 (s)		τ_R (ns)	
Proteins	Temperature (°C)	at 1.42 T	at 6.34 T	at 1.42 T	at 6.34 T
Myoglobin	36	0.030	0.40	18	16
Hemoglobin	36	0.062	0.75	47	30
Lysozyme	31	—	0.34	—	13
	43	0.031	0.21	19	8
Albumin	37	0.10	0.88	78	35

[a]From ref. 44 with permission.

molecule in which the C–H vectors concerned are embedded. The rotational motion of the entire molecule of the different proteins can be estimated roughly via the τ_R. On the other hand, the values of $\tau_R = 10$ ns and 44 ns are reported from dielectric relaxation measurements, respectively, for myoglobin[45] and hemoglobin,[46] and $\tau_R = 10$ ns from depolarized light scattering measurements for lysosyme,[47] although the source of proteins and the concentrations used in the measurements are somewhat different. It is encouraging to note that the values of τ_R obtained from quite different measurements are in rough agreement with each other.

The field dependence of τ_R can be seen in the table. In spherical myoglobin, the τ_R obtained at 1.42 and 6.34 T are in good mutual agreement. For hemoglobin, the difference between the two τ_R lies slightly outside the estimated experimental error. However, for the relatively nonspherical protein molecules lysozyme and albumin, the τ_R obtained at 1.42 T are about twice as long as those obtained at 6.34 T. The difference in τ_2 at the different field strengths for lysozyme and albumin suggests that the isotropic rigid motor model does not describe adequately the rotational motion of nonspherical proteins.

The analysis of T_1 data with the use of the axially symmetrical ellipsoid model was also conducted, changing the ratios of the long diameter to the short diameter. Nevertheless, it is reported[44] that the analysis does not provide significantly improved agreement among the correlation times (which in this case characterize the diffusional rotations about the long diameter axis as well as the short diameter axis separately) than was provided by the isotropic model. Presumably, internal motion of the C–H vectors in the ellipsoid also should be considered, whether the shape is spherical or nonspherical.

Such an attempt has been carried out by Howarth[10] Table 3-4 shows T_1 of the α-carbon envelope for various proteins at different field strengths and the analytical result obtained with the use of the spherical rotor model, including an internal librational motion of the C–H vector. Here, the values of τ_R(rigid) were calculated by eq. (3-73) on the basis of the spherical rigid rotor model. It is evident that the values of τ_R(rigid) depend strongly in most cases on the field strength and do not provide adequate insight into the rotational motion of the entire molecule. The values of τ_R(librational) were determined using eq. (3-101) with eqs. (3-102), (3-80), and (3-65) and $n = 1$, $r = 1.09A$, $\tau_R(\text{librational}) \equiv \tau_1$, $\Omega = 20°$ (or 0.35 rad) and $\tau_L = 10^{-11}$ s so as to render T_1 equal to the experimentally observed T_1 having the same value of τ_R(librational) at different field strengths for each protein. It is assumed that the C–H vector stochastically librates with the correlation time τ_L within the angle Ω about an axis in the assumed spherical molecule involving isotropic spherical random rotation with τ_R(librational). The values for Ω and τ_L were assumed empirically, and the librational rotational motion of the C–H vector (instead of stochastic rotation with a fixed angle to the axis) was introduced, considering the complex chain conformation of molecule. As can be seen, the calculated T_1 are in good agreement with the observed T_1 having an equivalent τ_R(librational) for each protein independent of the field strength. Hence, we can evaluate the overall

TABLE 3-4. FIELD DEPENDENCE OF T_1 FOR α-CARBON ENVELOPES OF PROTEINS AND ROTATIONAL CORRELATION TIMES CALCULATED FROM THE RIGID ROTOR MODEL AND LIBRATIONAL MODEL[a]

Protein	Concentration	pH	Temperature (°C)	Field (T)	T_1 (observed)[b]	τ_R (rigid)[c]	τ_R (scattering)[d]	τ_R (librational)[e]	T_1 (calculated)[b]
Sperm whale	16	7.5	32	1.41	0.036	25	—	11	0.037
cyanoferrimyoglobin	16	7.5	28	2.3	0.074	20	—	11	0.075
Bovine serum	3.0	6.5	38	1.42	0.10	78	77 (25°C)	40	0.08
albumin	3.2	5.1	31	4.2	0.57	49	77 (25°C)	40	0.55
	3.0	6.5	35	6.34	0.88	35	77 (25°C)	40	0.99
Lysozyme	15.4	3.1	31	6.34	0.34	13	10 (20°C)	9.5	0.34
	14.6	3	42	1.42	0.031	19	10 (20°C)	9.5	0.036
Ribonuclease A	2.3	6.6	30	1.41	0.035	22	—	8.0	0.036
	9.5	3.1	18	4.2	0.148	13		8.0	0.15
Alumichrome	150	DMSO	45	2.3	0.121	0.40	—	0.60	0.130
	150	DMSO	45	8.4	0.234	0.26		0.60	0.234
Gramicidin S	140	CD_3OD	43	1.41	0.141	0.35	—	0.52	0.138
	—	CD_3OD	42	6.34	0.194	0.32		0.52	0.206

[a]Data are taken from Table 1 of ref. 10.
[b]Carbon-13 T_1 for chemical shifts observable as an envelope of α carbons.
[c]Calculated from T_1 (obs) by Eq. (3-73) with $r = 1.09$ A, $n = 1$, $\tau_R = \tau_c$.
[d]Values determined from light-scattering experiments.
[e]Calculated by Eqs. (3-101) and (3-102) with $\Omega = 0.35$ rad ($= 20°$) and $\tau_L = 10^{-11}$ s. Here, $\tau_R = \tau_I$.

rotational motion of various proteins. It is also reported that such an analysis can be carried out for the protonated side-chain carbons of proteins by changing the angle of Ω. Consistent correlation times are obtained from the T_1 observed at different field strengths and from the nuclear Overhauser enhancements as well.

Motion of Side Chains in Polypeptides

Synthetic homopolypeptides provide useful models for studying chain conformations and dynamics in complex natural proteins. In most homopolypeptides, the chemical shifts of the individual carbons, ie, the main-chain α carbon and the side-chain carbons, are readily distinguished because of the simplicity of the chemical structure. Hence, examination of the spin relaxation provides detailed information about local chain conformations and dynamics. The rigid rotor model may be applicable to the rotational motion of the internuclear vector involving in the main chain to a first approximation. However, for ^{13}C spin relaxation of the side chains, additional internal motions of the C–H vectors must be considered. Because the chemical structure of the side chains is well defined for synthetic homopolypeptides, the validity of the ellipsoid rotational model can be judged by examining the spin relaxation phenomena in connection with the conformation and dynamics of the side chains. An example of this kind of study,[19,23,24] is the analysis made by Chachaty et al[24] of the ^{13}C T_1 data for poly[N^5-(3-hydroxypropyl)-L-glutamine] with the use of the ellipsoid rotor model involving multiple internal motions.

The chemical structure of the polymer is shown in Figure 3-8. It is first assumed that the C_α-H vector undergoes spherical random rotation because this polymer is found to be essentially in a coiled form in aqueous solution. However, a Cole-Cole distribution is introduced in eq. (3-110). The density function of Cole-Cole distribution is:

$$F(s) = \frac{1}{2\pi} \frac{\sin(\gamma\pi)}{\cosh(\gamma s) + \cos(\gamma\pi)} \tag{3-112}$$

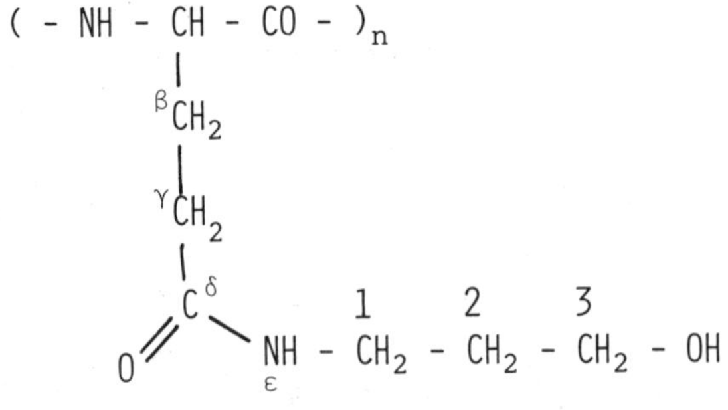

$$\begin{array}{l}
(-\mathrm{NH}-\overset{\alpha}{\mathrm{CH}}-\mathrm{CO}-)_n \\
\qquad\quad | \\
\qquad\quad {}^{\beta}\mathrm{CH_2} \\
\qquad\quad | \\
\qquad\quad {}^{\gamma}\mathrm{CH_2} \\
\qquad\quad | \\
\qquad\quad \mathrm{C}^{\delta} \qquad\quad 1 \qquad\; 2 \qquad\; 3 \\
\mathrm{O}\!=\!\!\!\!\quad\quad \diagdown\, \underset{\varepsilon}{\mathrm{NH}}-\mathrm{CH_2}-\mathrm{CH_2}-\mathrm{CH_2}-\mathrm{OH}
\end{array}$$

Figure 3-8. Chemical structure of poly[N^5-(3-hydroxypropyl)-L-glutamine].

where $s = \ln(\tau_c/\bar{\tau}_c)$, $\bar{\tau}_c$ being the center value of the distribution. Here, $0 < \gamma < 1$ is a parameter that determines the width of the distribution. Then, the spectral density is given by:

$$J_m(\omega) = K_m \frac{1}{\omega} \frac{\cos[(1-\gamma)(\pi/2)]}{\cosh(\gamma \ln \omega\bar{\tau}_c) + \sin[(1-\gamma)(\pi/2)]} \tag{3-113}$$

Examining the T_1 of the α carbon at 12–78°C and the field strengths of 2.1 and 5.9 T, Chachaty et al obtained the value of $\gamma = 0.7$ and $\bar{\tau}_c = \tau_0 \exp(\Delta H/RT)$ with $\tau_0 = 4.837 \times 10^{-14}$ s, $\Delta H = 6.0$ kcal mol^{-1}, where T is the temperature in degrees Kelvin. The individual C–H vectors in the side chain are assumed to rotate about the respective preceding bond vectors independently in different modes, ie, the C_β–H vector rotates about the C_α–C_β vector, the C_γ–H vector about the C_β–C_γ vector, the C_1–H vector about the C_γ–C_δ vector, the C_2–H vector about the C_1–C_2 vector, and the C_3–H vector about the C_2–C_3 vector. Here the rotation axis of the C_1–H vector is taken to be the C_γ–C_δ vector, as mentioned above, because if the rigid trans conformation is assumed for the $C_\gamma C_\delta$-N_ε-C_1 bond conformation, the vector N_ε–C_1 is roughly parallel to the C_γ–C_δ vector. The modes of rotation examined are: A, stochastic diffusional rotation; B, random jump rotation among three equivalent sites separated by 120° from each other; and C, random jump rotation among three sites, two of them being equivalent and symmetrical to the third site. In models B and C, the jump frequencies are defined to be W_1, W_2, and W_3, as illustrated in Figure 3-9.

The choice of adequate mode for each rotation was made by examining the

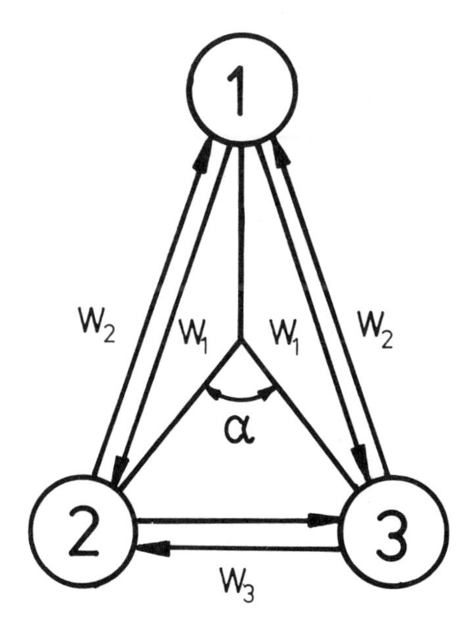

Figure 3-9. Illustration of random jump rotation among three sites and definition of jump frequencies.

TABLE 3-5. ANALYSIS OF T_1 DATA FOR SIDE-CHAIN CARBONS OF POLY [N^5-(3-HYDROXYPROPYL)-L-GLUTAMINE] OBSERVED AT 30°C FOR THE D_2O SOLUTION IN CONSIDERATION OF LOCAL ROTATIONAL MOTION OF EACH CARBON[a]

Nucleus	Field strength (T)	Rotation axis of C–H vector	Mode of rotation	Rotational jump rate (s^{-1})			Experiment T_1 (s)	Calculated T_1 (s)[b]
				W_1	W_2	W_3		
$^{13}C_\beta$	2.11	C_α–C_β	C	8×10^8	4.8×10^9	0	0.13	0.130
$^{13}C_\beta$	5.87	C_α–C_β	C			0	0.23	0.230
$^{13}C_\delta$	2.11	C_β–C_γ	C	1.4×10^9	7×10^8	0	0.19	0.196
$^{13}C_\gamma$	5.87	C_β–C_γ	C	1.4×10^9	7×10^8	0	0.27	0.272
$^{13}C_1$	2.11	C_γ–C_δ	A	3×10^9[c]			0.48	0.486
$^{13}C_1$	5.87	C_γ–C_δ	A	3×10^9[c]			0.54	0.538
$^{13}C_2$	2.11	C_1–C_2	B	4.5×10^9	4.5×10^9	4.5×10^9	0.99	0.994
$^{13}C_2$	5.87	C_1–C_2	B	4.5×10^9	4.5×10^9	4.5×10^9	0.105	0.103
$^{13}C_3$	2.11	C_2–C_3	B	3×10^9	3×10^9	3×10^9	0.15	0.150
$^{13}C_3$	5.87	C_2–C_3	B	3×10^9	3×10^9	3×10^9	0.15	0.151

[a] Taken from ref. 24 with permission.
[b] The calculation was made on the basis of the spherical random rotation of the C_α–H vector with the correlation time having a Cole-Cole distribution of $\gamma = 0.7$: $\bar{\tau} = 7.9 \times 10^{-10}$ s.
[c] Rotational diffusion constant (rad/s).

TABLE 3-6. OCCUPATION PROBABILITIES[a] OF ROTATION ISOMERS AT 30°C ESTIMATED BY ANALYSIS OF T_1

Rotation about bond	Occupation probabilities[b]		
	P_1	P_2	P_3
C_α–C_β	0.75 (0.63)	0.125 (0.24)	0.125 (0.13)
C_β–C_γ	0.20 (0.20)	0.40 (0.40)	0.40 (0.40)
C_1–C_2	0.333 (0.333)	0.333 (0.333)	0.333 (0.333)
C_2–C_3	0.333 (0.35)	0.333 (0.35)	0.333 (0.30)

[a]The values were calculated from W_1 and W_2 filed in Table 3-5 by Eqs. (3-114)[24].
[b]The values obtained from the proton-coupling constants are indicated in parentheses.

T_1 data obtained at two different field strengths and over a wide temperature range. It was pointed out that this choice chould be made exclusively for all rotations cited above so that only one of them could suffice to simulate the experimental data.

An example of analysis at 30°C is given in Table 3-5. As can be seen, the T_1 data obtained at two different field strengths are closely reproduced by adequate choice of the mode of each rotation and the jump rate. It is of interest here to estimate the populations of the rotational isomers, ie, the occupation probability at each site. The occupation probabilities P_1, P_2, and P_3, respectively, at sites 1, 2, and 3, can be evaluated by the transition rates W_1 and W_2 as:

$$P_1 = W_2/(2W_1 + W_2) \qquad \text{(3-114a)}$$

$$P_2 = P_3 = W_1/(2W_1 + W_2) \qquad \text{(3-114b)}$$

The values thus obtained from the data in Table 3-5 are summarized in Table 3-6 with the values obtained from the proton coupling constants. The occupation probabilities of the rotational isomers estimated from the two methods are mutually consistent. This consistency is significant, as it implies the validity of the modes of rotational motions assumed above and provides information regarding transition probabilities among the rotational isomers.

As shown by the above example, studies of spin relaxation of the side-chain residues of homopolypeptides or linear macromolecules provide insight into the chain dynamics and conformation insofar as the chemical structure is defined.

Spin Relaxation of Rubbery Polymers and Rubbery Components in Semicrystalline Polymers

The foregoing sections have concerned the spin relaxation of macromolecules in solution. Nevertheless, even for bulky macromolecular substances, high-resolution ^{13}C spectra can be obtained. The relaxation parameters can be evaluated with conventional spectrometers under a ^1H scalar decoupling condition if samples are in the rubbery state or involving rubbery material as a noncrystalline content,[33,34,48-52] because of the reduced ^{13}C–^1H dipolar interaction via rapid molecular motion. Nuclear magnetic resonance studies can be conducted only in a limited temperature range for polymers in solution; however, corresponding studies can be carried out for such bulk polymers over a wide range of temperature encompassing the temperature where the spin–lattice relaxation time T_1 indicates the minimum value, although the resolution of the spectrum is somewhat reduced because of the rapid decay of the transverse magnetization, ie, because of the very short transverse relaxation time T_2. Hence, the particular mode of chain motion of macromolecules can be detected by examining the temperature dependence of the relaxation parameter without changing the field strength. Furthermore, it is advantageous for this kind of study that macromolecular substances can be examined under conditions involving rather slow molecular motion. If the NMR work is conducted only in solution, the particular mode of molecular motion characteristic of very long-chain macromolecules may be hidden by the pronounced molecular motion and the insensibility of the relaxation parameters to the field strength.

Typical of this kind of study are extensive investigations of rubbery polymers, such as polyisoprene and polybutadiene, by Duch and Grant,[48] Schaefer,[34] and Komoroski et al[50] as well as of rubbery contents in semicrystalline polyethylene by Komoroski et al.[51] This section briefly reviews the work of this author and others of the noncrystalline rubbery components in crystalline terephthalic acid polyesters.[12] Figure 3-10 shows the ^{13}C NMR spectrum of poly(ethylene terephthalate) at different temperatures under a ^1H-decoupled condition. It is seen that the spectrum is very broad and almost structureless at the lowest temperature 115°C, but with increasing temperature resolvable lines that can be assigned to individual carbons progressively appear at the respective chemical shifts, which are in good agreement with those observed for the *o*-chlorophenol solution of this polymer. It is noted here that the lines of the terephthaloyl residue carbons C-1, C-2, and C-3 first appear, and that of C-4 adjacent to the ester linkage follows with increasing temperature. This indicates implicitly that the terephthaloyl residue first becomes mobile on the glass to rubber transformation. Until now, the glass to rubber transformation of polymers has been studied only in connection with the segmental motion of the overall molecule. It is of particular importance that ^{13}C NMR spectroscopy can provide insight into the contribution of individual constituent carbons to the transformation.

We have examined the ^{13}C NMR spectrum of a series of polyesters [-CO-⟨◯⟩-CO-O-$(CH_2)_n$-O-]$_x$, $n = 2, 3, 4, 6, 10$ over a wide range of temperature and found the following results:[12]

1. The individual carbons in the polyesters can be divided into three groups depending on their molecular mobility in a temperature range of the glass-rubber transformation: A, terephthaloyl carbons C-1, C-2, C-3; B,

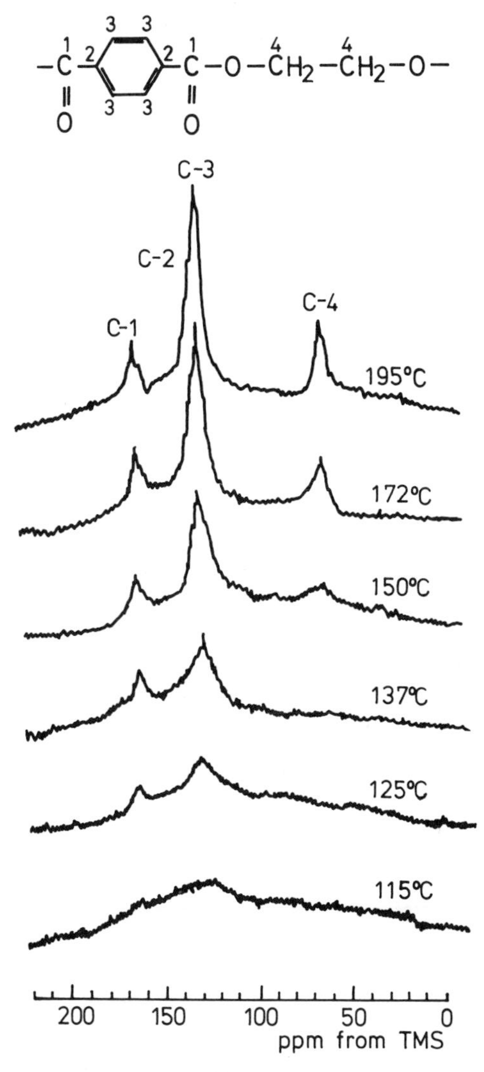

Figure 3-10. Proton scalar decoupled ^{13}C NMR of poly(ethylene terephthalate) at different temperatures.

terminal methylene carbon C-4 adjacent to the ester linkage; C, central methylene carbons C-5, C-6, etc.

2. For polyesters of $n = 2$ and 3 [poly(ethylene terephthalate) and poly(trimethylene terephthalate)], the terephthaloyl carbons are more mobile than are all methylene carbons.
3. For polyesters of $n \geqslant 4$, the central methylene carbons are most mobile, but the terminal methylene carbon is still highly restricted in motion.

This result is very significant because it suggests that local motions of the methylene sequence, such as the so-called three-bond motion illustrated in Figure 3-11, are able to take place, independent of terephthaloyl residues.

The detailed local motion of the individual carbons can be elucidated by analyzing the spin relaxation data with the aid of rotational models discussed under Theory of Magnetic Relaxation by Dipole–Dipole Interaction. We have measured the spin-lattice relaxation time T_1 for all terephthalic acid polyesters cited above over a wide range of temperature. Figure 3-12 exemplifies the result of poly(hexamethylene terephthalate). In this figure, nT_1 are plotted against the reciprocal of temperature in degrees Kelvin. Here, the temperature dependence of nT_1 can be seen over a much wider range than that in which T_1 measurements can be carried out in a solution of this polymer. Data have been obtained on T_1 in relation to the noncrystalline rubbery component of this crystalline polymer under conditions wherein the degree of segmental motion differs over a very wide range. Accordingly, these data provide particularly useful information regarding the detailed molecular motion characteristics of long-chain macromolecules.

As can be seen in the figure, the individual carbons exhibit distinctly different temperature dependencies of nT_1. The T_1 data are shown over a wide range of temperature including the melting temperature of this polymer. It should be noted first that no discontinuity can be seen beyond the melting zone (hatched zone in the figure) for all individual carbons, as reported also for the T_1–temperature plot of bulk polyethylene by Komoroski et al.[51] This fact provides evidence for the existence of noncrystalline material in the semicrystalline polymers, involving similar molecular conformation and molecular dynamics as in the molten state. Nevertheless, the point may be raised that one is, in fact, observing only a part of a noncrystalline material whose chain conformation is similar to that in the molten state. However, it is found that the ^{13}C T_1 observed for polyethylene by this technique under ^1H scalar decoupled

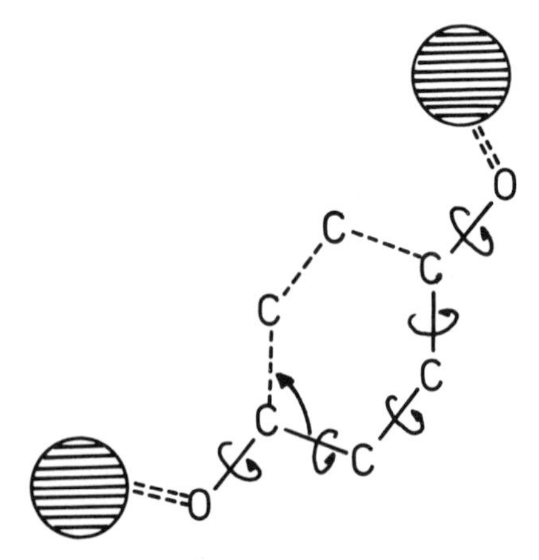

Figure 3-11. Schematic representation of three-bond motion available for methylene sequence in poly(tetramethylene terephthalate).

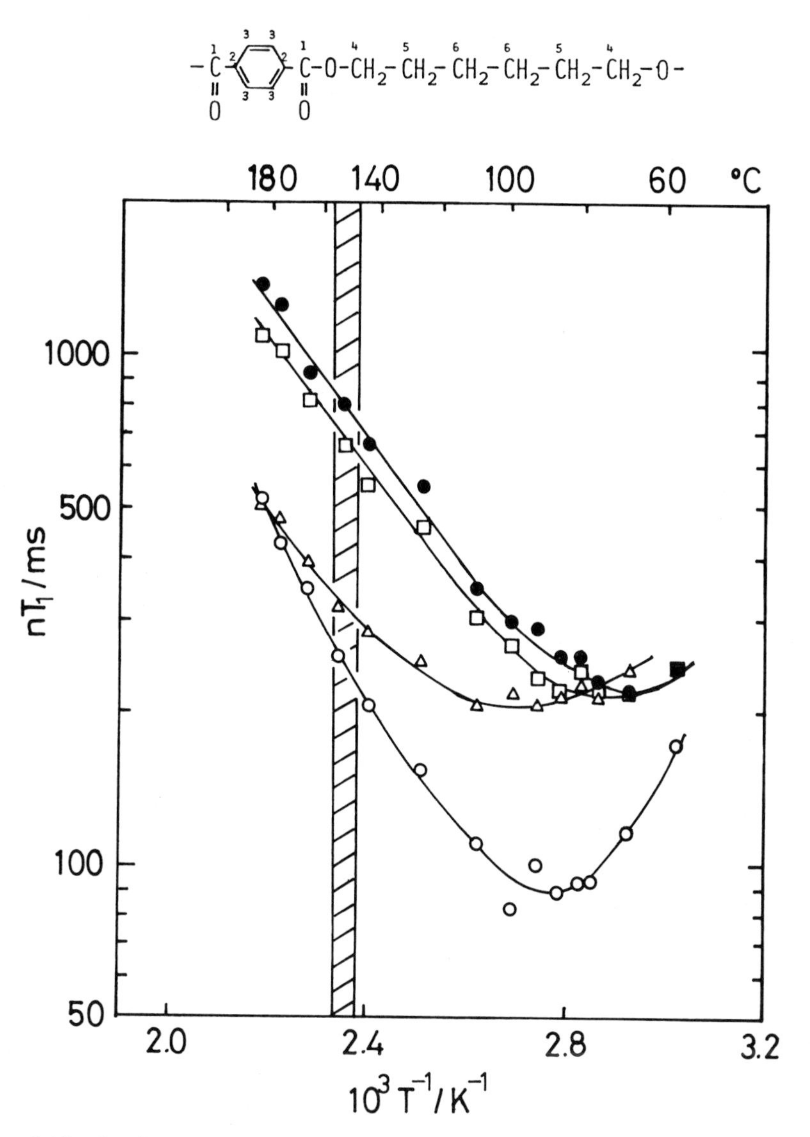

Figure 3-12. Semilogarithmic plots of ^{13}C nT_1 versus the reciprocal of absolute temperature for poly(hexamethylene terephthalate). ○, C-3; △, C-4; □, C-5; ●, C-6. The hatched zone indicates the melting temperature range determined by Differential Scanning Calorimetry (DSC).

conditions at room temperature coincide within the experimental error to the T_1 obtained by solid-state ^{13}C NMR (^1H dipole decoupled and magic angle sample spinning) (See Chapter 4), as ascribable to the entire noncrystalline content of this polymer.[52] Therefore, it must be concluded that the T_1 obtained here are ascribable to the entire noncrystalline component in the crystalline polymers. This conclusion also agrees with qualitative consideration of the intensity of the spectrum.

In the meantime, it is seen in the figure that the minimum values of nT_1 are about 200–210 ms for both of the terminal methylene carbon (C-4) and the inner methylene carbons (C-5 and C-6) and about 90 ms for the terephthaloyl methine carbon (C-3) at different ambient temperatures. The minimum values of nT_1 are not only different among the individual carbons but also much higher than those expected from the single correlation-time theory. Such high minimum values have not been explained by means of models that consider distributions of correlation times or by means of a defect diffusion model. Woessner's 2-τ model was also found to be inadequate. However, Howarth's 3-τ model in which the spectral density is described by eqs. (3-94) was found to be suitable to analyze the present nT_1 temperature data.[12,35]

In this model, three correlation times τ_I, τ_L, and τ_R and two angles Ω and Δ describe the modes of the rotational motions of the C–H vector as explained in the section, Molecular Motion with Plural Correlation Times, Pertinent to Flexible Long-Chain Macromolecules, by use of Figure 3-5. In this analysis, the correlation time τ_I is assumed to be common for all carbons but other parameters are assumed to be different for each carbon. Furthermore, it is assumed that τ_I and τ_L follow Arrhenius' temperature dependence, ie, $\tau_I = \tau_{IO} \exp(\Delta E_I/RT)$ and $\tau_L = \tau_{LO} \exp(\Delta E_L/RT)$, whereas τ_R is independent of temperature. Also, $\tau_R < \tau_L < \tau_I$ and $\Delta E_L < \Delta E_I$ are assumed in consideration of the intrinsic motion of polymeric chains. The vertical angles Ω and Δ, which characterize the modes of the libration and the rotational diffusions, are also assumed to be independent of temperature, although Ω may somewhat increase with increasing temperature.

Under these assumptions a trial and error analysis was carried out. All parameters were determined definitely. The trial and error analysis could be made satisfactorily despite the great number of parameters to be determined because some of the parameters involve a particular influence on the result; for example, the angle Δ primarily determines the minimum value of nT_1 and Ω the ambient temperature at which the minimum nT_1 appears. Sometimes, plural equivalent solutions were found to satisfy the experimental nT_1 behavior but it was not difficult to choose one of these, considering their physical meaning.

Table 3-7 summarizes the parameters thus determined for the individual carbons; the nT_1 vs temperature curves obtained using these parameters are shown as solid curves for two carbons (C-3 and C-6), as examples, in Figure 3-13. As can be seen, the calculated curves fit the experimental points closely for these protonated carbons. It is worth noting that the vertical angles, Δ, determined for the individual carbons without making any particular assumptions, are plausibly acceptable in consideration of the chain conformation. The values of Δ angles for all of the methylene carbons, ranging from 77 to 86°, are in approximate accordance with the supplementary angle (72°) of the bond angle C╲C╱H. Then, the angles thus determined reflect the fact that the C–H

TABLE 3-7. PARAMETERS OF 3-τ MODEL USED FOR THE CALCULATION OF THE nT_1 VALUES FOR PROTONATED CARBONS OF POLY(HEXAMETHYLENE TEREPHTHALATE)[a]

	Rotation		Libration			Isotropic motion	
Carbon	Δ (°)	τ_R (s)	Ω (°)	τ_{L0} (s)	ΔE_L (kJ mol^{-1})	τ_{I0} (s)	ΔE_I 9kJ mol^{-1})
C-3	29	1.0×10^{-11}	62	3.7×10^{-16}	49	1.1×10^{-15}	50
C-4	86	2.0×10^{-12}	39	9.7×10^{-15}	44		
C-5	80	2.0×10^{-12}	56	1.1×10^{-15}	43		
C-6	77	2.0×10^{-12}	58	1.1×10^{-15}	43		

[a]From ref. 12.

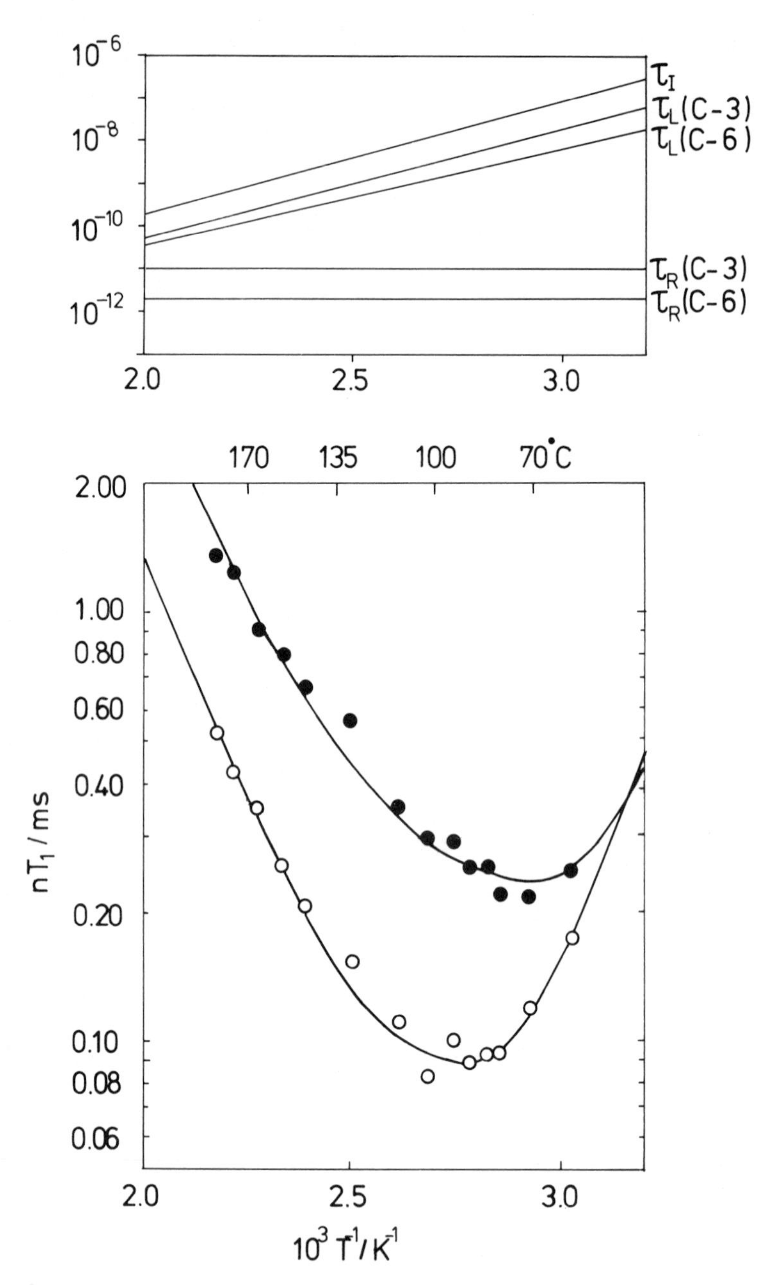

Figure 3-13. Analysis of T_1 data for C-3 and C-6 of poly(hexamethylene terephthalate) by the 3τ model. ○, C-3; ●, C-6. The parameters used are summarized in Table 3-7. τ_I and τ_L are plotted against the reciprocal of absolute temperature in the upper part.

vectors of methylene groups rotate around the C—C or O—C bonds. On the other hand, the Δ angle of 29° of the terephthaloyl methine carbon C-3 arises from the fact that the C–H vector can rotate by 30° around the long axis of the terephthaloyl residue, which is almost parallel to the direction of the methylene sequence in the trans,trans conformation. This axis, as well as those of the C—C and C—O bonds further librate within the individual cones, whose vertical angles Ω increase in the order of terminal methylene, central methylene, and aromatic methine groups. In Figure 3-12, τ_I, τ_L, and τ_R are also plotted against the reciprocal of temperature. The rates of the individual modes of motion involving in the individual carbons can readily be seen to be a function of temperature.

Proton scalar decoupled ^{13}C NMR spectroscopy, as exemplified above for a series of terephthalic acid polyesters, is generally applicable to studying the rubbery noncrystalline component of crystalline polymers. For example, the ^{13}C T_1 as well as T_2 of linear polyethylene samples crystallized either from the melt or from dilute solution were observed over a wide range of temperatures.[53,54] It was found that the T_1 were approximately equivalent for both samples, but T_2 were much shorter for the solution-grown sample than for the bulk-crystallized sample. This fact, together with the temperature dependencies of the spin relaxation times, could be understood only via analysis using the 3-τ model. It was concluded that the correlation time τ_I differed greatly between the two samples, but other parameters, such as τ_L and τ_R, relating to very local molecular motions, were equivalent for both samples. This conclusion is quite plausible in view of the morphological structure, depending on the mode of crystallization.

Relaxation Studies by Solid-State High-Resolution ^{13}C NMR

As described in Chapter 4 in detail,[56] high-resolution ^{13}C NMR can be conducted also for solid substances using techniques, combining cross polarization (CP), ^{1}H dipolar decoupling (DD), and magic-angle sample spinning (MASS). Solid-state ^{13}C NMR can be applied to the study of many macromolecular substances that contain only ^{13}C and ^{1}H as nonzero spin nuclei. These substances can be studied in the bulk state, not only in the rubbery state but also in the glassy or crystalline state.

In this case, the relaxation of the longitudinal ^{13}C magnetization is also theoretically expected to follow eq. (3-3) as discussed in solving the coupled differential eqs. (3-20) if the relaxation is conducted under conditions where the ^{1}H longitudinal magnetization stays zero or constant. As described in most textbooks, eq. (3-3) is valid in the extreme case that $\rho=\rho'=2\delta$ because the coupled differential eqs. (3-20) could be easily solved to give eq. (3-3). However, as discussed in the section, Carbon-13 Spin Relaxation in α Carbons in Proteins, in solving eqs. (3-20), eq. (3-3) is also valid in solid-state ^{13}C NMR if ^{1}H longitudinal magnetization is zero or stays constant during relaxation. In solid-state ^{13}C NMR, the high power of H^1 dipolar decoupling is, of course, applied during signal acquisition so that the ^1H magnetization can be considered zero. However, ^1H decoupling is not always applied during periods other than signal acquisition. Therefore, one must be careful to choose conditions such that the above-mentioned condition is fulfilled; otherwise there is no theoretical basis to the assumption that ^{13}C longitudinal relaxation follows eq. (3-3).

Therefore, the spin–lattice relaxation time T_1 can be measured for macromolecular substances in the bulk state using appropriate pulse programs. Particularly, it is advantageous in solid-state ^{13}C NMR to be able to detect relaxation in connection with the anisotropy of the chemical shift. For example, if the NMR experiment is conducted for uniaxially oriented crystalline polymers lying so that the molecular chain axis is parallel to the static magnetization field without MASS, the chemical shift corresponding to the 3-3 component (σ_{33}) of the chemical shift tensor is observed.[56] In such an experiment using uniaxially oriented polyethylene, the σ_{33} element of the chemical shift tensor corresponding the trans,trans conformation of this polymer was indeed observed with the σ_{iso} ascribable to the noncrystalline component and coinciding with the chemical shift of this polymer in solution or in the melt.[57,58] The T_1 of these σ_{33} and σ_{iso} are considered in relation to the molecular mobility of the crystalline and noncrystalline components.

On the other hand, if the NMR experiment is conducted for isotropic samples with magic-angle sample spinning, the σ_{ave} (the average value of the diagonal elements of the chemical shift tensor) for a particular chain conformation of polymers is observable.[55] In such work with DD–MASS on crystalline polyethylene, σ_{ave} ascribable to the trans,trans chain conformation and σ_{iso} were detected[52,56,58-60] and the respective T_1 were examined.[52,59,60] As the result, it is found that σ_{ave} consists of two components associated with different T_1, one of which is on the order of several seconds and the other several hundred seconds, whereas σ_{iso} contains only one component.[52,59,60] The identification of T_1 of several seconds for the crystalline σ_{ave} is of particular importance because it suggests the existence of a molecular mobility whose correlation time is on the order of 10^{-7} to 10^{-8} s within the crystalline region.[52]

The T_1 as well as the mass fractions of the crystalline and noncrystalline components corresponding to different T_1 depend on the detailed sample morphology. Table 3-8 summarizes the T_1 and the mass fractions for two samples of linear polyethylene with different crystallization conditions, one crystallized from the melt (bulk crystal) and the other crystallized from dilute solution.[52,13] It is worth noting that a sizeable fraction of the crystalline material (the mass fraction of 0.59) is associated with a molecular motion in the solution-grown sample, whereas only a small amount of crystalline material (the mass fraction of 0.14) is associated with molecular mobility in the bulk-crystal sample. This difference between the two samples is particularly significant with regard to morphological differences. Further study of the mode of the molecular motion is expected.

TABLE 3-8. THE CHEMICAL SHIFTS AND T_1 BY SOLID-STATE ^{13}C NMR ANALYSIS FOR POLYETHYLENE SAMPLES CRYSTALLIZED FROM THE MELT (BULK CRYSTAL) OR DILUTE SOLUTION (SOLUTION CRYSTAL)[a]

Sample	Component	Chemical shift (ppm)[b]	Mass fraction	T_1 (s)
Bulk crystal	Rigid crystalline	34 (σ_{ave})	0.47	>300
	Mobile crystalline	34 (σ_{ave})	0.14	1.6
	Noncrystalline	32 (σ_{iso})	0.39	0.23
Solution crystal	Rigid crystalline	34 (σ_{ave})	0.19	≃200
	Mobile crystalline	34 (σ_{ave})	0.59	2.0
	Noncrystalline	32 (σ_{iso})	0.22	0.23

[a]The mass fraction of the total crystalline component was first determined by the solid-state ^{13}C spectrum by the pulse program $(\pi/4 - 1000s)_n$ without cross polarization and the mobile crystalline component was next determined by the pulse program $(\pi/4 - 10s)$ eliminating the contribution from the rigid crystalline component (the longest T_1 component).

[b]The chemical shift is indicated in ppm based on that of tetramethylsilane.

References

1. Bloemberger, N.; Purcell, E. M.; Pound, R. V. *Phys. Rev.* **1948**, *73*, 679.
2. Kubo, R.; Tomita, K. *J. Phys. Soc. Jpn.* **1954**, *9*, 888.
3. Solomon, I. *Phys. Rev.* **1955**, *99*, 559.
4. Abragam, A. "The Principle of Nuclear Magnetism"; Oxford Univ. Press: Oxford, 1961.
5. Slichter, C. P. "Principles of Magnetic Resonance"; Springer-Verlag: Heidelberg, 1978.
6. Dixon, W. T. "Theory and Interpretation of Magnetic Resonance Spectra"; Plenum: New York, 1972.
7. See Chapter 8 of ref. 4.
8. Woessner, D. E. *J. Chem. Phys.* **1962**, *37*, 647.
9. Woessner, D. E. *J. Chem. Phys.* **1962**, *36*, 1.
10. Howarth, O. W. *J. Chem. Soc., Faraday Trans. II* **1978**, *74*, 1031.
11. Howarth, O. W. *J. Chem. Soc., Faraday Trans. II* **1979**, *75*, 863.
12. Horii, F.; Hirai, A.; Murayama, K.; Kitamaru, R.; Suzuki, T. *Macromolecules* **1983**, *16*, 273.
13. Kitamaru, R.; Horii, F.; Murayama, K. "Abstracts of Papers", page 647, 28th Macromolecular Symposium, IUPAC, Amherst, Mass., July 12–16, 1982.
14. Kitamaru, R. "Abstract of Papers", 14th Europhysics Conference on Macromolecular Physics, European Physical Society, Vilafranca del Penedes, Spain, September 21–24, 1982.
15. Murayama, K.; Horii, F.; Kitamaru, R. *Bull. Inst. Chem. Res., Kyoto Univ.*, **1983**, *61*, 229.
16. Gronski, W.; Murayama, N. *Macromol. Chem.* **1978**, *197*, 1521.
17. Jones, A. A. *J. Polym. Sci., Polym. Phys. Ed.* **1977**, *15*, 863.
18. Ghesquiere, D.; BuuBan; Chachaty, C. *Macromolecules* **1979**, *12*, 775.
19. Tsutsumi, A. *Mol. Phys.* **1979**, *37*, 111.
20. London, R. E.; Avitabile, J. *J. Am. Chem. Soc.* **1977**, *99*, 7765.
21. Tsutsumi, A.; Chachaty, C. *Macromolecules* **1979**, *12*, 429.
22. Huntress, W. T. *J. Chem. Phys.* **1968**, *48*, 3524.
23. Tsutsumi, A.; Quaegebeur, J. P.; Chachaty, C. *Mol. Phys.* **1979**, *38*, 1717.
24. Perly, B.; Chachaty, C.; Tsutsumi, A. *J. Am. Chem. Soc.* **1980**, *102*, 1521.
25. de Gennes, P.-G. "Scaling Concepts in Polymer Physics"; Cornell Univ. Press: Ithaca, NY, 1979.
26. Doi, M.; Edwards, S. F. *J. Chem. Soc., Faraday Trans. II* **1978**, *74*, 1789, 1802, 1818.
27. Kimmich, R. *Polymer* **1977**, *18*, 233.
28. Koch, H.; Bachus, R.; Kimmich, R. *Polymer* **1980**, *21*, 1009.
29. Howarth, O. W. *J. Chem. Soc., Faraday Trans. II* **1980**, *76*, 1219.

30. Levine, Y. K.; Birdsall, N. J. M.; Lee, A. G.; Metcalfe, J. C.; Partington, P.; Roberts, G. C. K. *J. Chem. Phys.* **1974**, *60*, 2890.
31. Connor, T. M. *Trans. Faraday Soc.* **1964**, *60*, 1574.
32. Heatley, F.; Begun, A. *Polymer* **1976**, *17*, 399.
33. Schaefer, J. *Macromolecules* **1973**, *6*, 882.
34. Schaefer, J. *Macromolecules* **1972**, *5*, 427.
35. Murayama, K.; Hirai, A.; Horii, F.; Kitamaru, R. *Polym. Prepr.* (Jpn) **1981**, *30*, 2006.
36. Hama, T.; Suzuki, T.; Kosaka, K. *Kobunshi Ronbunshu* **1975**, *32*, 91.
37. Axelson, D. E.; Mandelkern, L.; Levy, G. C. *Macromolecules* **1977**, *10*, 557.
38. Inoue, Y.; Nishioka, A.; Chujo, R. *J. Polym. Sci., Polym. Phys. Ed.* **1973**, *11*, 2237.
39. Schaefer, J.; Natusch, D. F. S. *Macromolecules* **1972**, *5*, 416.
40. Kuhlmann, K. F.; Grant, D. M.; Harris, R. K. *J. Chem. Phys.* **1970**, *52*, 3439.
41. Allerhand, A.; Doddrell, D.; Glushko, V.; Cochran, D. W.; Wenkert, E.; Lawson, J.; Gurd, F. R. N. *J. Am. Chem. Soc.* **1971**, *93*, 544.
42. Allerhand, A.; Doddrell, D.; Komoroski, R. *J. Chem. Phys.* **1971**, *55*, 189.
43. Norton, R. S.; Clouse, A. O.; Addleman, R.; Allerhand, A. *J. Am. Chem. Soc.* **1977**, *99*, 79.
44. Wilbur, D. J.; Norton, R. S.; Clouse, A. O.; Addleman, R.; Allerhand, A. *J. Am. Chem. Soc.* **1976**, *98*, 8250.
45. Marcy, H. O.; Wyman, J. *J. Am. Chem. Soc.* **1942**, *64*, 638.
46. Schlecht, P.; Vogel, H.; Mayer, A. *Biopolymers* **1968**, *6*, 1717.
47. Dubin, S. B.; Clark, N. A.; Benedek, G. B. *J. Chem. Phys.* **1971**, *54*, 5158.
48. Duch, M. W.; Grant, D. M. *Macromolecules* **1970**, *3*, 1965.
49. Schaefer, J.; Chin, S. H.; Weissman, S. I. *Macromolecules* **1972**, **5**, 798.
50. Komoroski, R. A.; Maxfield, J.; Mandelkern, L. *Macromolecules* **1977**, *10*, 545.
51. Komoroski, R. A.; Maxfield, J.; Sakaguchi, F.; Mandelkern, L. *Macromolecules* **1977**, *10*, 550.
52. Kitamaru, R.; Horii, F.; Murayama, K. *Polym. Bull.* **1982**, *7*, 583.
53. Horii, F.; Murayama, K.; Kitamaru, R. *Polym. Prepr. ACS* **1983**, *24*, 384.
54. Mandelkern, L. *Pure Appl. Chem.* **1982**, *54*, 611.
55. See, for example, Mehring, M. "High Resolution NMR in Solids", Springer-Verlag: Berlin, 1983.
56. Earl, W. L.; VanderHart, D. L. *Macromolecules* **1979**, *12*, 762.
57. Kitamaru, R.; Horii, F.; Terao, T.; Maeda, S.; Saika, A. *Ann. Rep. Res. Inst. Chem. Fibers, Jpn.* **1981**, *38*, 49.; Horii, F.; Kitamaru, R.; Maeda, S.; Saika, A.; Terao, T. *Polym. Bull.* **1985**, *13*, 179.
58. Fyfe, C. A.; Lyerla, J. R.; Volksen, W.; Yannoni, C. S. *Macromolecules* **1979**, *12*, 757.
59. Schröter, B.; Posern, A. *Makromol. Chem.* **1981**, *182*, 675.
60. Schröter, B.; Posern, A. *Makromol. Chem., Rapid Commun.* **1982**, *3*, 623.

4

STEREOCHEMICAL STUDIES IN THE SOLID STATE

Takehiko Terao and Fumio Imashiro

Department of Chemistry
Faculty of Science
Kyoto University
Kyoto 606, Japan

Introduction

To date, NMR studies of stereochemistry in the solid state have been made by measurements of broad-line spectra and/or relaxation times mainly of abundant spins. During the last several years, however, high-resolution solid-state NMR for dilute spins has become very popular and provided detailed and interesting information. In this chapter stereochemical studies in the solid state that have been performed mainly by high-resolution solid-state NMR for dilute spins are considered. The first part gives an outline of high-resolution solid-state NMR for dilute spins. The second part demonstrates that determination of ^{13}C chemical shift anisotropy in combination with ^{1}H relaxation time measurements is an extremely useful method for investigating molecular motions in solids. The last part reviews applications of ^{13}C cross-polarization magic-angle spinning (CPMAS) NMR to stereochemical studies in the solid state.

High-Resolution Solid-State ^{13}C NMR

In this section, after the general theory of high-resolution NMR for dilute spins in solids is given, recent advances in this method and experimental means for motional studies using it are reviewed briefly.

NMR in Stereochemical Analysis

Principles

The monograph written by M. Mehring[1] is highly recommended for an understanding of principles of high-resolution solid-state NMR. Here, only a brief explanation is given.

In a conventional ^{13}C powder spectrum, the extremely broad linewidth caused by ^{13}C–^{1}H dipolar interactions and ^{13}C chemical shift anisotropies obscures the fine structure resulting from isotropic chemical shifts. In liquids, both interactions are averaged randomly to zero in ordinary coordinate space by rapid molecular motion. In order to obtain high-resolution spectra in solids, the ^{13}C–^{1}H dipolar interactions must be removed artificially. These dipolar interactions may be averaged coherently to zero in spin space by rapidly inverting ^{1}H spins with a strong radiofrequency (rf) field ($\sim$15 G). In single crystals, chemical shift anisotropy does not cause line broadening but makes the chemical shift angular dependent. From the chemical-shift rotation pattern measured with high-power ^{1}H decoupling, complete information regarding the shift tensors can be obtained. Measurement of a chemical shift tensor allows interpretation of the local static and dynamic environment of the nucleus in three dimensions. When a powder sample is used for measurements, the spectrum of each functional group of a molecule shows a characteristic shape caused by chemical shift anisotropy under strong proton decoupling. In favorable cases, the analysis of lineshape allows us to obtain the principal values of the shift tensor for each functional group, but not the principal axes; in general, however, the lineshape is too complex to be interpreted. The linewidth from chemical shift anisotropy can be removed by rotating a sample around the axis, making the magic angle of 54.7° with the static field; this technique is called magic-angle spinning (MAS). The combination of heteronuclear dipolar decoupling and MAS provides a means of obtaining high-resolution isotropic spectra of dilute spins in solids.

However, one serious problem still remains: in addition to the relatively small magnetic moment and low natural abundance (1.1%) of ^{13}C, the ^{13}C spin–lattice relaxation time T_1 is extremely long in many solids, so that a normal Fourier transform (FT) NMR is not practical. This problem can be solved by using a method called cross polarization (CP). In the state where the proton spin system is locked along an rf field $H_{1\mathrm{H}}$, another rf field $H_{1\mathrm{C}}$ is applied to the ^{13}C spin system under the Hartman-Hahn condition defined by the equation $\gamma_{\mathrm{H}} H_{1\mathrm{H}} = \gamma_{\mathrm{C}} H_{1\mathrm{C}}$, where γ_{H} and γ_{C} are the magnetogyric ratios of ^{1}H and ^{13}C, respectively. Then energy exchange takes place between the ^{1}H and ^{13}C spin systems via the ^{13}C–^{1}H dipolar coupling. When the whole system has achieved a thermal equilibrium state, the ^{13}C polarization—enhanced by a factor of four in comparison with a thermal equilibrium magnetization of the ^{13}C spin system—is obtained along the $H_{1\mathrm{C}}$ direction. The CP method therefore provides a large magnetization in short time compared with a normal FT NMR. The magnetization generated along the $H_{1\mathrm{C}}$ direction is observed as a free induction decay (FID) by turning off $H_{1\mathrm{C}}$. A FID that yields a high-resolution spectrum by Fourier transform can be obtained with H_{1H} left on for ^{13}C–^{1}H dipolar decoupling. The combined techniques of high-power decoupling, CP, and MAS are often called CPMAS NMR spectroscopy.

Recent Developments

In addition to simply observing spectra by the above-mentioned method, useful information can be obtained by using various versions of solid-state ^{13}C NMR spectroscopy summarized in this section, although some of them are still being developed.

The ^{13}C–^{1}H dipolar coupling tensor, which can be obtained only in solids, permits determination of internuclear distances, and, moreover, provides a means of studying molecular motions. The separation of chemical shifts and ^{13}C–^{1}H dipolar interactions in the NMR spectra of single crystals can be accomplished[2-4] with the aid of two-dimensional (2D) spectroscopy.[5,6] This method is also applicable to powder samples.[7,8] However, when overlap from chemically nonequivalent nuclei occurs, the spectra are difficult to interpret; the combination of this experiment and MAS may remove the overlap.[9-14] The chemical shift dipolar 2D spectrum also provides information on the mutual orientation of the two tensors, even in powder samples. An alternative 2D NMR technique[15] has been proposed, which is one version of switching-angle sample-spinning (SASS) spectroscopy.[16] The transverse magnetization is developed at an off-magic angle under homonuclear decoupling and detected at the magic angle under heteronuclear decoupling. Carbon–carbon bond lengths in amorphous solids can be measured by observing the nutation signals.[17,18]

Chemical shift anisotropy $\Delta\sigma$ is another useful parameter that is available only in solids, reflecting details of the local electronic environment and molecular motion. In many cases, however, a normal spectrum for a nonspinning powder sample shows extensive overlap of the signals originating from different sites in the molecule. Several strategies have been proposed to overcome this problem. One type involves observation of spinning sidebands,[19] which flank an isotropic centerband at integral multiplets of the spinning frequency Ω when $\Omega \lesssim \Delta\sigma$. The anisotropy powder patterns can be reconstructed from the spinning sidebands in several ways.[19-21] Another type of strategy consists of applying a train of rf pulses synchronized to the revolution of the spinner.[22-24] The application of such a pulse train causes the refocusing tendency of MAS to be offset, allowing reconstruction of the powder patterns. Individual powder patterns of chemically nonequivalent carbons in a molecule can be obtained separately by rotating a sample in three discrete steps[25] or by another version[16,26] of SASS spectroscopy.

The indirect spin–spin coupling constant J is an important parameter reflecting electronic environment and also can be used as a tool of resonance assignments. In normal CPMAS NMR, however, heteronuclear decoupling also eliminates heteronuclear indirect spin–spin coupling at the same time. This difficulty has been removed by using homonuclear rather than heteronuclear decoupling.[27] Under homonuclear dipolar decoupling, MAS can remove the inhomogeneous heteronuclear dipolar interaction, leaving heteronuclear indirect spin–spin coupling. Analogs[28,29] of CPMAS for the attached proton test and heteronuclear 2D *J*-resolved NMR spectroscopy[15,30] can be performed by modifying this method. Selective observation of nonprotonated carbons is possible by the modified technique;[28,29] a simpler method[31] is available, based on the fact that ^{13}C–^{1}H dipolar interactions of nonprotonated carbons are particularly weak.

Two-dimensional NMR techniques for performing heteronuclear chemical shift correlation experiments in single crystals[32] and in rotating solids[33,34] have been reported. The techniques simplify assignment and, moreover, substantially improve the overall proton resolution *vis-à-vis* conventional methods because of the large spectral dispersions and superior resolution in the ^{13}C dimension of the spectra.

When the shift anisotropy is not of interest, spectra without sidebands are desired; sidebands decrease the sensitivity and make not only quantitative analyses but also signal assignments quite difficult. A novel method[35,36] for solving all of these problems has been proposed in which a stable distribution of rotating-frame magnetization is established to yield an FID without rotary echoes by coaddition of echo signals.

Motional Studies

Studies of molecular motions in solids using ^{13}C NMR spectroscopy can be made by measurements of relaxation times and/or spectral lineshapes, as is the case with other types of NMR techniques. ^{13}C relaxation time measurements have an advantage of selectivity compared with the case of protons; the scheme for ^{13}C spin–lattice relaxation time T_{1C} in the laboratory frame[37] or $T_{1\rho C}$ in the rotating frame[38] can be carried out easily. Measurements of ^{13}C relaxation times, however, have some limitations: T_{1C} measurements in solids require too long a time, except in the neighborhood of T_{1C} minima, and the observed $T_{1\rho C}$ value reflects molecular motion only when it is much shorter than the spin–spin cross-polarization time characteristic of the rigid lattice, because the spin–spin coupling between the spin-locked state and the dipolar reservoir can be a powerful relaxation mechanism.[39]

The powder pattern of chemical shift anisotropy displays especially characteristic features if the molecule undergoes a rotational jump motion with the jump frequency being of the same order of magnitude as the width of the spectrum in the rigid case. For this frequency region, lineshape analysis provides a rather direct means, not only of determining the jump frequency but also of distinguishing between types of motion.[40,41]

An NMR technique for characterizing slower motions (10^3 to 10^{-2} reorientations per second) is available.[42] In this method holes can be burned into a powder spectrum, tagging molecules that are at certain orientations. Subsequent molecular reorientations brings about a spectral diffusion. By measuring the broadening and recovery of the hole as a function of time, detailed knowledge of the reorientations is obtained (eg, correlation time and angular jump size).

The presence of incoherent averaging (molecular motion) reduces the efficiency of coherent averaging resulting from rf decoupling or MAS and leads to relaxation of the S spins, when the former averaging rate is comparable to that of the latter.[43,44] This fact provides a way to study slow motions, which has an advantage in that the relaxation times are determined on the basis of a single spectrum. To analyze the experimental data, however, a model for the molecular motion must be assumed.

Additionally, molecular motions can be investigated by observing dipolar interactions between ^{13}C and ^{1}H spins[9-15] or ^{13}C and ^{14}N spins.[45] Chemical exchange processes may be studied by a CPMAS analog of 2D exchange spectroscopy.[46] An investigation of molecular motion by ^{13}C NMR should be made by using the above-mentioned methods properly, according to the correlation time, whether or not a model for the motion is known, etc.

Studies of Molecular Motions: ^{13}C Spectrum and Proton T_1 Measurements

The powder pattern from chemical shift anisotropy is quite sensitive to the type and the correlation time of the molecular motion: however, individual powder patterns must be distinguishable from others in order to perform a detailed study of the motion. Moreover, the correlation time can be determined only when it is considerably long. Therefore, conventional, simple proton T_1 measurements are still useful, and, in combination with ^{13}C high-resolution NMR without MAS, they can be a powerful tool for studying molecular motions. Using this combination, we have studied (1) dynamic behavior of methanol molecules trapped in β-quinol clathrate,[47,48] (2) proton transfer in the benzoic acid dimer,[49] and (3) in-plane molecular rotation of naphthalenes.[50] We discuss our results in this section.

Dynamic Behavior of Methanol Molecules Trapped in β-Quinol Clathrate

In β-quinol methanol clathrate, one methanol molecule is trapped, not chemically but physically, in a cavity constructed by six quinol molecules, referred to as the host.[51,52] Showing typical molecular motions in molecular crystals, β-quinol clathrate is one of the most suitable systems for a detailed study of molecular motion. The purpose of this investigation[47,48] is to clarify the behavior of guest methanol molecules over a wide temperature range.

The preferred orientations of the guest methanol molecule are shown in Figure 4-1.[42] These consist of a set of three orientations related by C_3 symmetry about the c axis and another set of three orientations related by inversion of the first three through the cavity center. x-Ray crystallography cannot determine whether each CO bond axis remains in one of the six directions or reorients itself among them. In ^{13}C high-resolution NMR, however, if the former is the case, three lines must be observed for the guest molecules, and if the C_3 reorientation exists, they should coalesce to a single line. We observed 10 lines in general crystal orientations at room temperature. Nine lines are expected for quinol molecules from the crystal structure:[51] each quinol molecule in β-quinol clathrate possesses a center of symmetry, and there are three magnetically nonequivalent molecules in the unit cell. Therefore, only one line can be assigned to methanol molecules. This provides evidence for the existence of the C_3 reorientation of each CO bond axis at room temperature. Because two methanol molecules related by inversion have the same chemical shift tensor, the existence of interchange between the two sets of three CO orientations can be neither affirmed nor denied using high-resolution NMR. However, the existence of such interchange has been shown by dielectric measurements;[53] this interchange affects proton T_1.

The 10 ^{13}C chemical shift tensors were determined from measurement of the rotation pattern; only the tensor of the guest methanol molecules is discussed here. The tensor is axially symmetrical around the c axis, reflecting the C_3 reorientation.

Figure 4-2 shows the temperature dependence of the chemical shift anisotropy and the isotropic mean shift for the guest methanol molecule. It is known that this substance has a transition point at 67 K. Above the transition point the

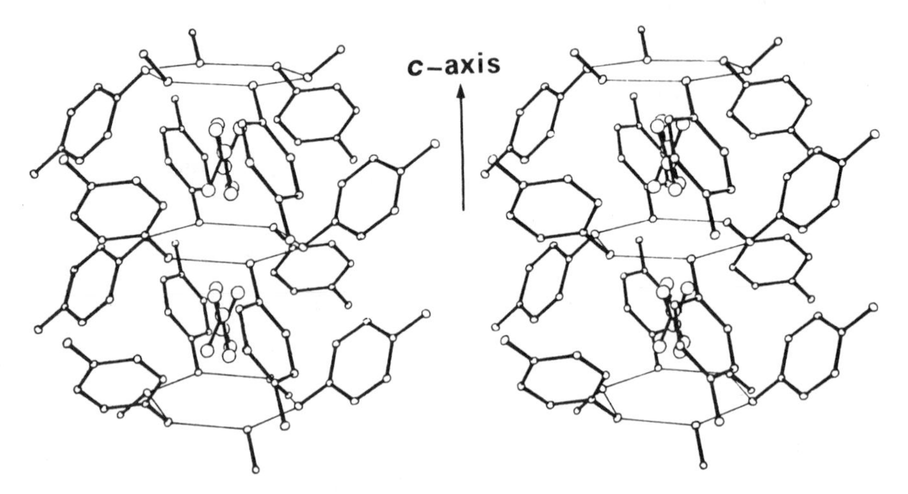

Figure 4-1. Stereoview of β-quinol methanol clathrate: open and closed circles in cavities denote oxygen and carbon atoms of methanol molecules, respectively. The six possible positions for the CO axis are indicated.

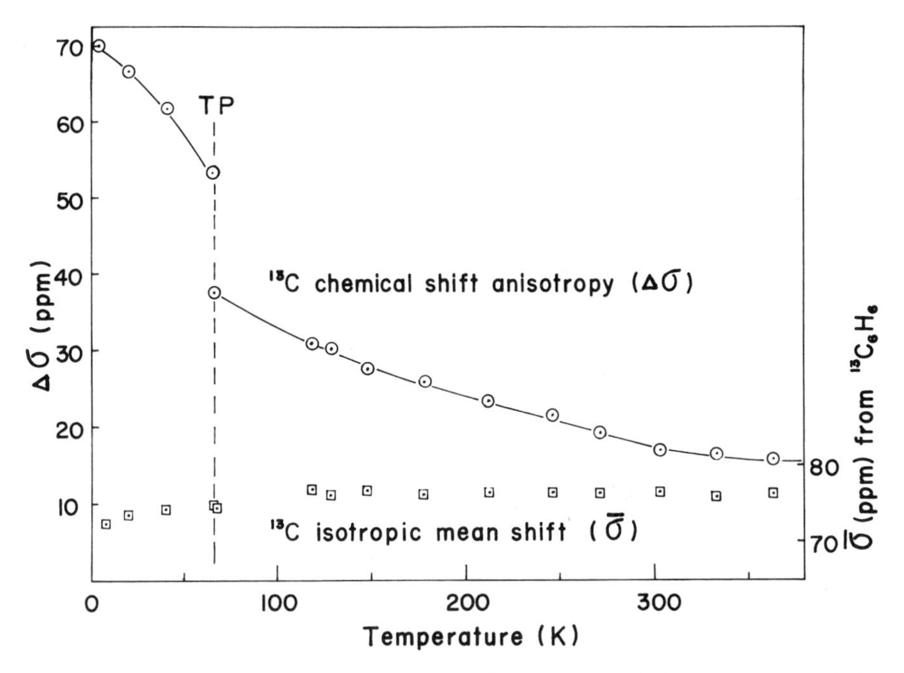

Figure 4-2. Temperature dependence of the ^{13}C chemical shift anisotropy and the isotropic mean shift of the guest methanol. (Reproduced with permission from ref. 48.)

tensor remains axially symmetrical about the c axis. The anisotropy increases gradually with decrease in temperature; it is considerably smaller than the value (68 ppm) obtained for polycrystalline methanol throughout the high-temperature phase. This is caused by scaling down of the anisotropy by the C_3 reorientation of the methanol molecule with its CO bond axis making an angle θ with the c axis. Neglecting the deviation from axial symmetry and taking account of the geometrical symmetry around the carbon, it can be reasonably assumed that the unique axis of the ^{13}C chemical shift tensor for a methanol molecule is parallel to the CO bond direction. Under this assumption, the chemical shift anisotropy $\Delta\sigma_{obs}$ for the C_3 reorienting methanol is expressed in terms of the anisotropy $\Delta\sigma_0$ for the static molecule and the angle θ as:

$$\Delta\sigma_{obs} = P_2(\cos\theta)\Delta\sigma_0 \tag{4-1}$$

Using 67.8 ppm as the value of $\Delta\sigma_0$ which we determined for polycrystalline methanol, the temperature dependence of the angle θ can be obtained from the temperature-dependent anisotropy shown in Figure 4-2; the angle decreases continuously from 46 to 33° with decrease of temperature from 363 to 67 K.

Figure 4-3 shows the temperature dependence of the ^{13}C spectrum of the guest molecule at a fixed crystal orientation; a sample with 20% ^{13}C-enriched methanol molecules was used for obtaining the signals of the methanol carbons distinctively from those of the other carbons. Although a single line is observed above the transition point, it abruptly splits into three when temperature is

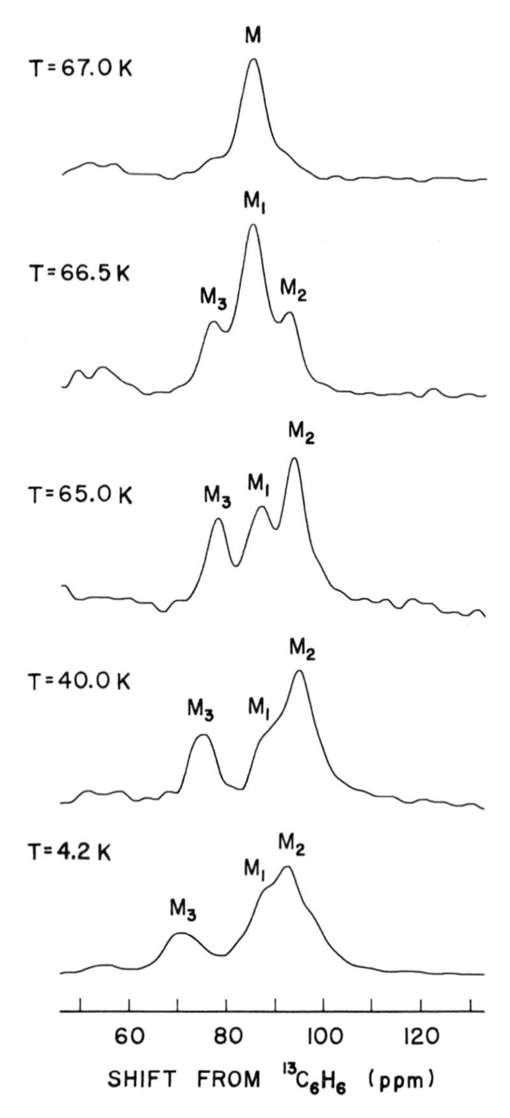

Figure 4-3. Temperature dependence of the ^{13}C high-resolution spectrum of the guest methanol at a fixed crystal orientation. (Reproduced with permission from ref. 48.)

reduced through the transition point. This splitting directly shows quenching of the C_3 orientation of the guest molecules below the transition point. The line positions, however, are still temperature dependent, suggesting a change of the orientations of the methanol molecules and/or the appearance of another type of motion. In order to obtain more detailed information, we determined the chemical shift tensors at 65, 40, 20, and 4.2 K. Assuming that the most shielded direction and the CO bond axis are parallel, the angle θ between the CO bond and the c axis is determined to be 25°, 29°, 30°, and 31° at 65, 40, 20, and 4.2 K, respectively. Also, assuming that the intermediate shielded direction is perpendicular to the COH plane,[54,55] the angle ψ between the OH bond direction and the plane parallel to the CO bond axis and the c axis may be determined to be 16°, 12°, 13°, and 3° at 65, 40, 20, and 4.2 K, respectively: the COH plane comes closer to the c axis with decrease in temperature.

The three distinct sets of principal axis directions are not equivalent under the C_3 symmetry operations, the deviations being 3–6° and over the experimental error of $\pm 1°$. Therefore, the guest–host interaction potential is believed to lose the C_3 symmetry below the transition point, implying the possibility of a slight ordering of the component of the methanol dipole moment perpendicular to the c axis.

Reflecting quenching of the C_3 reorientation, the anisotropy abruptly increases when temperature is reduced through the transition point, as seen in Figure 4-2. However, the anisotropy is still smaller just below the transition point than the value of 67.8 ppm for polycrystalline methanol, indicating the existence of some motion. We believe that libration of the guest molecules takes place, accompanying quenching of the C_3 reorientation. The increase of anisotropy with decrease in temperature results from the decrease of librational amplitude. This decrease brings about a corresponding increase in the angle θ between the CO bond and the c axis. If it is assumed that each methanol molecule undergoes harmonic rotational vibrations with an amplitude Θ around the two orthogonal axes perpendicular to the CO bond, one of which is parallel to the COH plane, then the librational effect on the anisotropy is given[56] by:

$$\Delta\sigma_{\text{lib}} = \left[1 - \frac{3h}{8\pi^2 I\nu}\coth\left(\frac{h\nu}{2kT}\right)\right]\Delta\sigma_{\text{st}} \tag{4-2}$$

where $\Delta\sigma_{\text{lib}}$ denotes the observed anisotropy and $\Delta\sigma_{\text{st}}$ the anisotropy in the absence of libration, ν is the librational frequency, and I is the molecular moment of inertia averaged with respect to the two libration axes. The least-squares fitting of eq. (4-2) to the experimental values in Figure 4-2 gives 76.0 ± 1.0 ppm for the anisotropy $\Delta\sigma_{\text{st}}$ and 26.6 ± 1.0 cm^{-1} for the librational frequency ν. Then the librational amplitude in degree is given by:

$$(\langle\Theta^2\rangle)^{1/2} = \left(\frac{h}{8\pi^2 I\nu}\right)^{1/2}\coth^{1/2}\left(\frac{h\nu}{2kT}\right) = 9.86\coth^{1/2}(19.1/T) \tag{4-3}$$

In this way detailed information was obtained on the behavior of the guest methanol molecules trapped in β-quinol clathrate using ^{13}C high-resolution solid-state NMR. However, the motional parameters for the C_3 reorientation and the interchange between the two sets of three C_3 orientations cannot be given by this method. Accordingly, the temperature dependence of proton T_1 was measured in β-quinol clathrates with CH_3OH and CD_3OH trapped as the guest at 60 MHz mainly for obtaining the activation energies and the correlation times: the results are shown in Figure 4-4. Each minimum observed in the high-temperature phase arises from the interchange process. The slope appearing just above the transition point is attributable to the C_3 reorientation. The minima in the low-temperature phase result from the tunneling of the methyl group. In the case of the CD_3OH guest sample, it seems likely that the minimum should be related to the tunneling of the CH_2D group in its natural abundance.[57,58]

The abrupt change at the transition point shows that the C_3 reorientation of most guest molecules disappears as the temperature falls through the transition point. However, a small portion of the guest molecules may still undergo the C_3 reorientation even below the transition point, the portion decreasing rapidly with decrease in temperature. This is reflected in the temperature dependence of T_1 just below the transition point, and in the methanol ^{13}C spectrum at 66.5 K as well, where the intensity of the central line containing a signal from reorienting molecules is apparently larger than those of the other lines. Such residual phenomena may be characteristic of clathrate, because all the cavities are not completely filled with guest molecules.

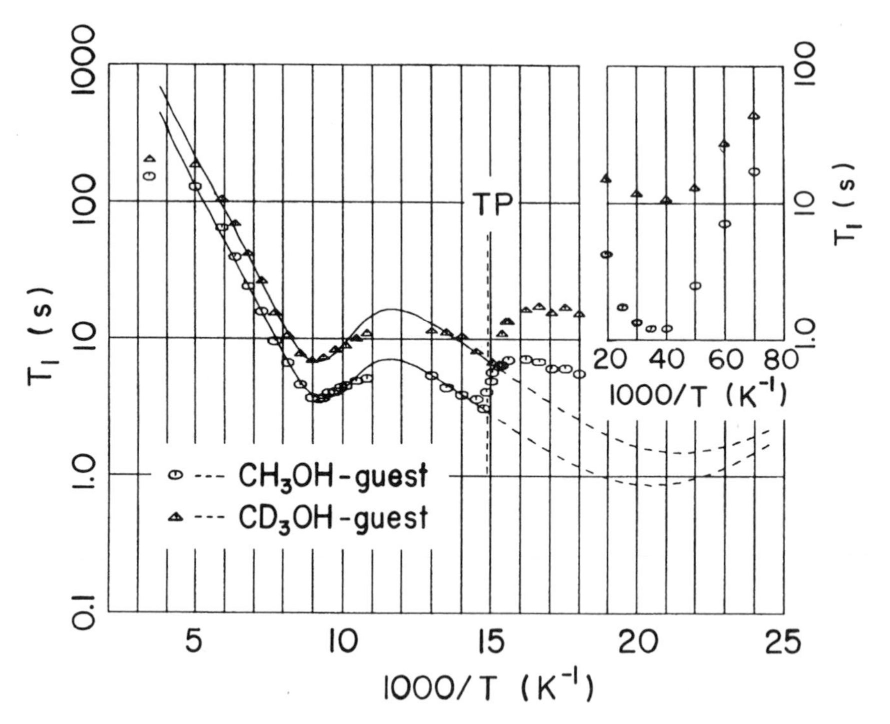

Figure 4-4. Temperature dependence of the ^1H spin–lattice relaxation times in β-quinol clathrates with CH_3OH and CD_3OH trapped as the guest. The solid lines show the theoretical fits. The broken lines show theoretical curves in the fictitious case in which the phase transition and the methyl tunneling are absent. (Reproduced with permission from ref. 48.)

The proton T_1 formula for the coexisting two types of motions above the transition piont was derived. According to the general formulation,[59-61] T_1 can be expressed by:

$$\frac{1}{T_1} = \sum_{l=1}^{3} C_l B(\tau_l) \qquad (4\text{-}4)$$

where C_l and τ_l are the structure-dependent coefficient and the correlation time for the lth normal mode of reorientation, respectively, and $B(\tau)$ is given by

$$B(\tau) = \frac{\tau}{1 + \omega^2\tau^2} + \frac{4\tau}{1 + 4\omega^2\tau^2} \qquad (4\text{-}5)$$

The correlation time τ is assumed to be given by the Arrhenius expression:

$$\tau = \tau_0 \exp(E_a/RT) \qquad (4\text{-}6)$$

where E_a is the activation energy and τ_0 is the correlation time at infinite temperature. The coordinates of all the protons of the host and the guest are necessary to calculate C_3; we determined the OH orientation and position of the methanol molecule in the cavity from the T_1 minimum values in the high-temperature phase. With these results and the angle θ obtained by ^{13}C NMR, we calculated the temperature dependence of T_1. The theoretical fits to the experimental points give the motional parameters: the activation energies and the correlation times at $T_1 = \infty$ are 0.7 kcal mol^{-1} and 3.7×10^{-12} s, respectively, for the C_3 reorientation, and 2.4 kcal mol^{-1} and 4.5×10^{-14} s, respectively, for the interchange process. Thus, details of the behavior of the guest methanol molecules trapped in β-quinol clathrate over a wide temperature range have been elucidated using a combination of ^{13}C chemical shift tensor measurements and proton T_1 experiments.

Proton Transfer in the Benzoic Acid Dimer

It has been suggested by infrared (IR) studies[62,63] that in benzoic acid (BAC) two different configurations (A and B shown in Figure 4-5a) coexist in the solid state. These configurations are interconverted by simultaneous proton transfer along the two hydrogen bonds of the dimer. Whereas the two configurations are energetically degenerate in the free hydrogen bonded dimers, the degeneracy is removed in the solid state by distortion of the double-minimum intramolecular potential from intermolecular interactions. We measured the temperature dependence of ^{13}C high-resolution spectra in a single crystal and of proton T_1 to obtain detailed information about the two configurations; direct evidence for proton transfer, the conversion rate and potential barrier for the transfer, and the relative population of the two configurations were thereby determined.

The crystal structure of BAC is monoclinic with the space group $P2_1/c$.[64] The unit cell contains four molecules that form two magnetically nonequivalent dimers; each dimer possesses the two configurations A and B.

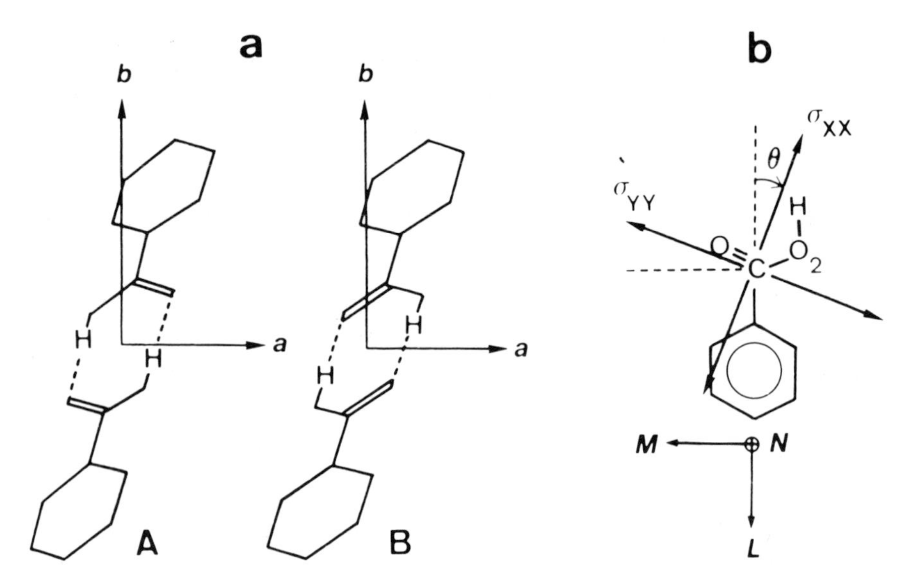

Figure 4-5. (a) Projection of BAC molecules onto the *ac* plane in configurations A and B. (b) The directions of the principal axes (X, Y, Z) of the chemical shift tensor and the molecular fixed axes (L, M, N).

Direct evidence for proton transfer can be obtained in the same way as mentioned previously for C_3 reorientation of the guest methanol molecules in β-quinol clathrate. If the two configurations coexist not dynamically but statically, four ^{13}C lines should be observed for the carboxyl carbons in general crystal orientations; in fact, we observed only two ^{13}C lines. This observation definitely indicates the existence of proton transfer with the conversion rate rapid enough to average out the differences in the line positions of the two configurations.

We measured the temperature dependence of the ^{13}C chemical shift tensors ($\boldsymbol{\sigma}$) of the carboxyl carbons using carboxyl ^{13}C 5%-enriched single crystals. The directions of the principal axes (X, Y, Z) were found as shown in Figure 4-5b; the Z axes are parallel to the N axis of the molecular frame (L, M, N) within $\pm 3°$. The chemical shift parameters including the angle θ defined in Figure 4-5b are summarized in Table 4-1, showing that only σ_{XX}, σ_{YY}, and θ are dependent on temperature. Accordingly, we analyzed the results in the following way: because configuration A can be converted to configuration B and vice versa by reflection with respect to the LN plane, the ^{13}C chemical shift tensors $\boldsymbol{\sigma}_A$ and $\boldsymbol{\sigma}_B$ in the LMN axis system for the molecules in configurations A and B, respectively, are written as

$$\boldsymbol{\sigma}_A = \begin{bmatrix} \sigma_{LL} & \sigma_{LM} & 0 \\ \sigma_{LM} & \sigma_{MM} & 0 \\ 0 & 0 & \sigma_{NN} \end{bmatrix}, \quad \boldsymbol{\sigma}_B = \begin{bmatrix} \sigma_{LL} & -\sigma_{LM} & 0 \\ -\sigma_{LM} & -\sigma_{MM} & 0 \\ 0 & 0 & \sigma_{NN} \end{bmatrix} \qquad (4\text{-}7)$$

TABLE 4-1. Chemical Shift Tensors of Carboxyl Carbon in Benzoic Acid[a]

T (K)	σ_{XX}	σ_{YY}	σ_{ZZ}	$\frac{1}{3}\mathrm{Tr}\boldsymbol{\sigma}$	θ[b]	P_A[c]
298	227	186	107	173	6	0.54
152	231	186	103	173	13	0.58
77	233	183	102	173	22	0.66
55	238	180	103	174	23	0.71

[a]In ppm from tetramethylsilane.
[b]θ is defined in Figure 4-5b.
[c]P_A denotes the probability of finding configuration A.

Here the Z axes are taken to be parallel to the N axis. The observed chemical shift tensor ($\boldsymbol{\sigma}_{obs}$) is given by the weighted average of $\boldsymbol{\sigma}_A$ and $\boldsymbol{\sigma}_B$:

$$\boldsymbol{\sigma}_{obs} = P_A\boldsymbol{\sigma}_A + P_B\boldsymbol{\sigma}_B \tag{4-8}$$

Here P_A (P_B) denotes the probability of finding configuration A (B). We assume that the energy difference between the two configurations does not depend upon the temperature and there is no entropy difference between the two configurations. Then P_A (P_B) is given by

$$P_A = \frac{a}{a+1}\left(P_B = \frac{1}{a+1}\right), \tag{4-9}$$

where $a = \exp(\Delta E/RT)$.

We obtained $\sigma_{LM} = -51.5$ ppm and $\Delta E = 0.10 \pm 0.01$ kcal mol^{-1} from the temperature dependence of the LM element of $\boldsymbol{\sigma}_{obs}$ given by $[(a-1)/(a+1)]\sigma_{LM}$. The ΔE obtained is in excellent agreement with the value determined by Hayashi and Kimura[63] from the temperature dependence of the IR spectra.

The angle θ defined in Figure 4-5b is determined to be 35° for the configuration existing at 0 K. This value of θ indicates that the Y direction is nearly along the C=O direction if A is the lower energy configuration, and that both X and Y directions substantially deviate from the C=O direction if B is the lower energy one. It is known that the direction of one of the in-plane principal axes of the shielding tensor is directed nearly along the C=O direction in molecules containing the C=O groups, such as esters and ketones.[55, 65-67] Therefore, it is concluded that configuration A is the lower energy configuration. The P_A values calculated from the obtained ΔE are listed in Table 4-1.

In order to obtain further insight into the proton-transfer dynamics, proton T_1 measurements for BAC and BAC-d_5 (ring protons deuterated) were made at temperatures between 15 and 370 K. Using the spin–lattice relaxation theory for classical jumps between configurations with unequal potential depth,[59-61] T_1 for the BAC dimer is given by:

$$\frac{1}{T_1} = CB(\tau)\frac{a}{(1+a)^2} \tag{4-10}$$

where

$$C = \frac{3}{5N}\gamma^4\hbar^2 \sum_i \left(\frac{1}{r_{iA}^6} + \frac{1}{r_{iB}^6} + \frac{1 - 3\cos^2\theta_{iAB}}{r_{iA}^3 r_{iB}^3} \right) \tag{4-11}$$

and

$$\tau = \tau_0 \frac{\exp(V_0/RT)}{1+a} \tag{4-12}$$

Here r_{iA} (r_{iB}) denotes the distance between the protons of the ith pair in configuration A (B) and θ_{iAB} is the angle between the proton pairs in the two configurations; N is the number of protons in the dimer (12 in BAC, two in BAC-d_5); V_0 is the height of the potential barrier above the lower potential minimum. In the limit $\omega\tau \ll 1$, eq. (4-10) reduces to

$$\frac{\ln T_1}{(1+a)^3} = \frac{-(V_0 + \Delta E)}{RT} - \ln 5C\tau_0 \tag{4-13}$$

Plots of ln T_1 versus $1/T$ for BAC and BAC-d_5 exhibit straight lines with equal negative gradients between 120 and 370 K. using $\Delta E = 0.10$ kcal mol^{-1}, $V_0 = 1.17 \pm 0.02$ kcal mol^{-1} is obtained from the gradients. This potential depth is much smaller than that estimated theoretically for a formic acid dimer;[68-70] however, geometrical optimization produces values that are much closer to the experimental value.[71-73]

Equation (4-10), with the geometry given in ref. 74, $V_0 = 1.17$ kcal mol^{-1}, and $\Delta E = 0.10$ kcal mol^{-1}, gives T_1(BAC)/T_1(BAC-d_5) = 4.2, when dipole–dipole interactions between all protons within 3.5 Å around the carboxylic hydrogens are taken into account. The fact that the slopes for BAC and BAC-d_5 are the same and the calculated T_1(BAC)/T_1(BAC-d_5) ratio is in fair agreement with the observed value of 3.8 insures that proton transfer along the hydrogen bonds is responsible for the proton relaxation.

The rates, $k_{A\to B}$ and $k_{B\to A}$, of proton transfer between configurations A and B are estimated to be:

$$\begin{aligned} k_{A\to B} &= 1/\tau_0 \exp(V_0/RT) \\ &= 2.34 \times 10^{11} \exp(5.89 \times 10^2/T)\ \mathrm{s}^{-1} \\ k_{B\to A} &= k_{A\to B}a = k_{A\to B} \exp(5.03 \times 10/T)\ \mathrm{s}^{-1} \end{aligned} \tag{4-14}$$

Here τ_0 is obtained from the extrapolated value of T_1 in the limit $T \to \infty$ and C is evaluated from the geometry given in ref. 74.

Below 120 K plots of ln T_1 versus $1/T$ deviate markedly from the behavior predicted by the relaxation theory for classical jumps. This result indicates that proton transfer via a tunneling mechanism becomes important at temperatures lower than 120 K.

Similar results have been reported for p-toluic acid by measurements of proton dipolar spectra and proton T_1,[75] and benzoic acid derivatives and decanoic acid by proton T_1 measurements.[76]

In-Plane Molecular Rotation of Naphthalenes

In-plane molecular rotation is often observed even in the solid state for aromatic molecules with a (near) symmetry axis perpendicular to the molecular plane, when the intermolecular potential barrier to rotation is not appreciable. In NMR studies of substituted benzenes[77-87] it is concluded that the barriers to the in-plane molecular rotation are lower in more disklike molecules,[88] and also lower in molecules substituted by a chlorine atom rather than a methyl group, the volume of which is similar to a chlorine atom.[89] This conclusion holds well also for large aromatic molecules: the activation energy of the in-plane rotation is much larger for pyrene (13.5 kcal mol^{-1} for reorientation between equivalent sites)[90] and for azulene (15.4 kcal mol^{-1} for 180° flipping)[91] than for coronene (5.8 kcal mol^{-1} for reorientation around the sixfold axis).[92] Pyrene and azulene molecules are rather elliptical, whereas a coronene molecule is very large but disklike. Therefore, it is not surprising that the in-plane molecular rotation of naphthalene molecules takes place only just below the melting point with a very large activation energy (25 ± 2 kcal mol^{-1}),[93] and that the flat, elongated anthracene molecule does not undergo any overall molecular motion.[94] Recently, however, we have found that some substituted naphthalene molecules with no symmetry axes and nondisklike shapes undergo overall molecular motion with notably small activation energies.[50] The molecular motion is confirmed to be from the in-plane molecular rotation about the axis perpendicular to the naphthalene ring by measurements of ^{13}C powder-pattern NMR spectra and the temperature dependence of ^{1}H second moments.

In the proton-decoupled ^{13}C NMR spectra for powder samples each resonance signal of a carbon atom appears as a powder pattern. When a fast uniaxial molecular rotation occurs, each powder pattern becomes axially symmetrical, permitting determination of the rotation axis direction relative to the shielding axis system. Further, under a rapid isotropic overall molecular rotation, the ^{13}C resonance signal is observed as a sharp single line at the isotropic mean position of the three principal values, which agrees with the resonance position measured by the ^{13}C CPMAS NMR method. On the other hand, a 180° flipping about any axis does not affect a powder pattern because chemical shift tensors for the present naphthalenes were not yet known. information about molecular motion from a powder spectrum, the chemical shift tensor for the stationary molecule must be known, however, the ^{13}C chemical shift tensors for the present naphthalenes were not yet known. Accordingly, they were estimated from those reported for methyl-substituted benzenes:[95-101] on the average, $\sigma_{11} = 224$ ppm, $\sigma_{22} = 134$ ppm, and $\sigma_{33} = 24$ ppm for the protonated carbons, although the scatter of σ_{33} is appreciable, and $\sigma_{11} = 225$ ppm, $\sigma_{22} = 160$ ppm, and $\sigma_{33} = 20$ ppm for the methyl-substituted carbons. The principal values for the representative ^{13}C chemical shift tensor of the present naphthalenes therefore are taken to be 220, 150, and 20 ppm for σ_{11}, σ_{22}, and σ_{33}, respectively, and the principal axis of σ_{33} is considered to be perpendicular to the naphthalene ring as found for the benzenes.[1] The calculated powder-pattern spectrum for this chemical tensor is shown in Figure 4-6a. If the correlation time for in-plane molecular rotation is much shorter than 10^{-4} s, the resonance signal observed at 15 MHz appears as an axially symmetrical powder pattern with $\sigma_{\parallel} = (1/2)(\sigma_{11} + \sigma_{22}) = 185$ ppm and $\sigma_{\perp} = \sigma_{33} = 20$ ppm (Figure 4-6b). If the correlation time for the isotropic overall molecular motions is much shorter than 5×10^{-5} s, the ^{13}C resonance line centers at $\sigma_{iso} = (1/3)\mathrm{Tr}\boldsymbol{\sigma} = 130$ ppm (Figure 4-6c).

In Figure 4-7 are shown the ^{13}C powder-pattern NMR spectra measured at room temperature for the three peri-methyl-substituted naphthalenes: 1,8-dimethylnaphthalene (**1**), 1,4,5,8-tetramethylnaphthalene (**2**), and 1-methyl-8-bromonaphthalene (**3**). The aromatic regions for **1** and **2** consist of many peaks spread over a wide range (20–230 ppm), whereas the aromatic region for **3** consists of a superposition of axially symmetrical powder-pattern signals with an absorption maximum at 185 ppm. This value is identical with $\sigma_{\parallel}$ in Figure 4-6b. Consequently, the powder-pattern spectra observed for **1** and **2** are thought to consist of many asymmetrical powder patterns, indicating the absence of molecular motion except for the possibility of a 180° flipping; on the other

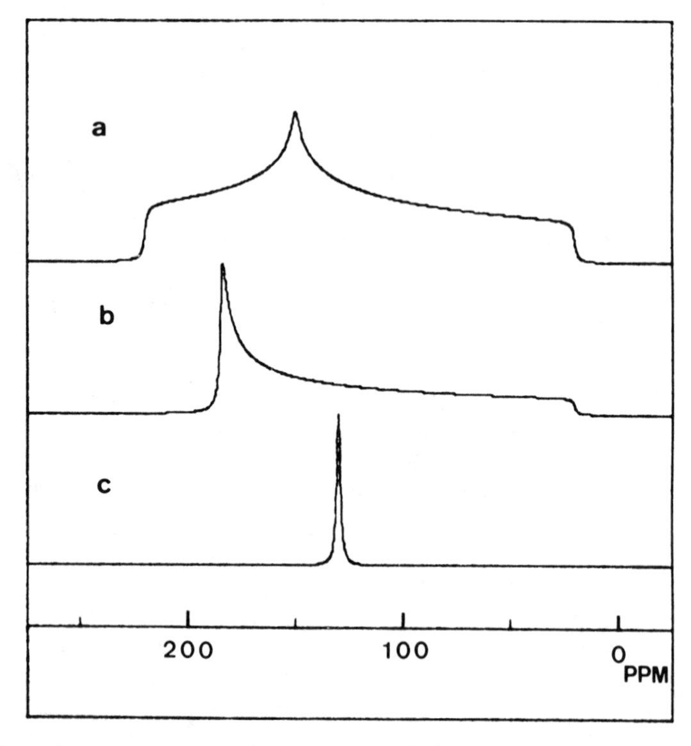

Figure 4-6. Theoretical powder lineshapes convoluted with Lorentzian broadening functions. The linewidth is 1 ppm. (a) $\sigma_{11} = 220$ ppm, $\sigma_{22} = 150$ ppm, and $\sigma_{33} = 20$ ppm; (b) $\sigma_{\parallel} = 185$ ppm and $\sigma_{\perp} = 20$ ppm; (c) $\sigma_{iso} = 130$ ppm.

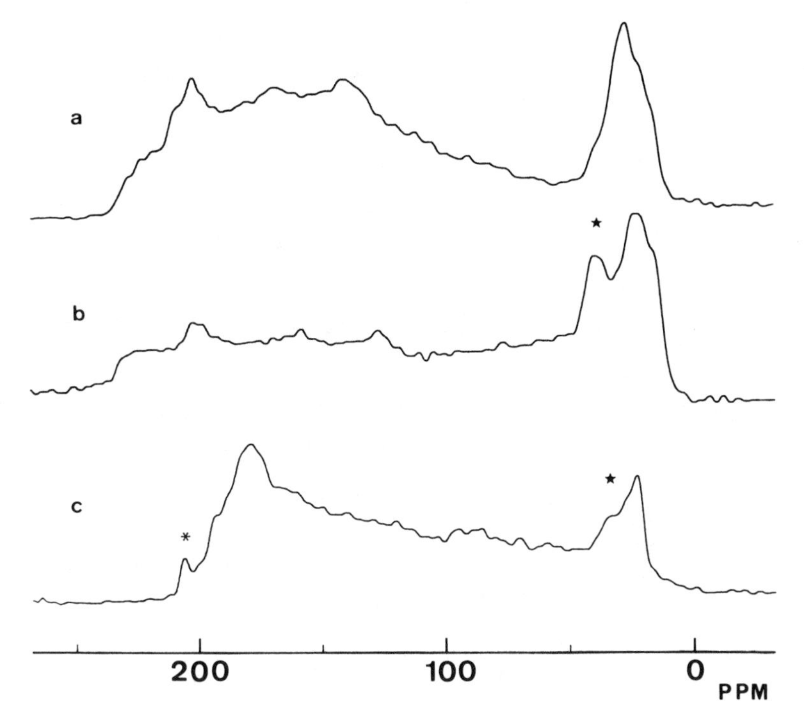

Figure 4-7. Carbon-13 powder-pattern NMR spectra of peri-methyl-substituted naphthalenes at room temperature. (a) 1,8-dimethylnaphthalene **1**. (b) 1,4,5,8-tetramethylnaphthalene **2**. (c) 1-methyl-8-bromonaphthalene **3**. The peak marked with asterisk denotes a spurious noise. The origin of the peaks marked with stars is unknown. (Reproduced with permission after ref. 50.)

hand, the spectrum for **3** is well interpreted by assuming the presence of the fast in-plane molecular rotation.

CH_3 CH_3 CH_3 CH_3 Br CH_3

CH_3 CH_3

1   **2**   **3**

This in-plane molecular rotation reduces proton–proton dipolar interactions, which are evaluated by the ^{1}H second moment. The second moment observed for **3** declines smoothly from 15.9 G^2 at 77 K to a plateau value of 1.3 G^2 above ~250 K.[50] Because the crystal structure of **3** was not yet available, the second moment was calculated by assuming[102] that **3** and **1** were isomorphous. The calculated rigid-lattice value (15.4 G^2) is in good agreement with the experimental value (15.9 G^2) at 77 K, and the high-temperature plateau value (1.3 G^2) is close to the value (1.7 G^2) calculated for a combination of the methyl rotation and the in-plane molecular rotation.

The parameters for molecular motion are determined by analysis of the temperature dependence of the ^{1}H spin–lattice relaxation time, T_1. The experimental results are shown in Figure 4-8. Equations (4-4)–(4-6) describe T_1. Three types of motions are considered: (1) the C_3 reorientation of the methyl groups and (2) the in-plane 180° flipping, and (3) rotational diffusion for overall molecular motion.

For random thermal reorientation of the methyl group, C is given by:[103-106]

$$C = N_{\mathrm{Me}} \frac{2}{N_{\mathrm{all}}} \frac{9}{40} \frac{\gamma^4 \hbar^2}{r^6} \tag{4-15}$$

where r is the intramethyl proton–proton distance, N_{all} is the total number of

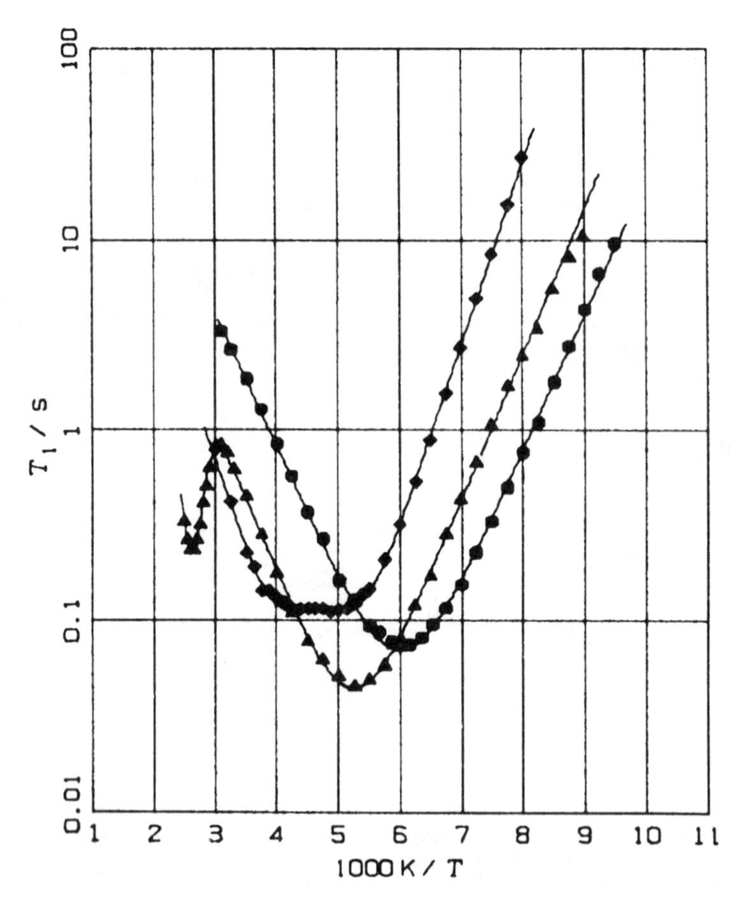

Figure 4-8. Temperature dependence of T_1 for 1,8-dimethylnaphthalene (**1**: filled circles), 1,4,5,8-tetramethylnaphthalene (**2**: filled triangles), and 1-methyl-8-bromonaphthalene (**3**: filled squares). (Reproduced with permission after ref. 50.)

protons in the molecule, and N_{Me} is the number of protons in the methyl group, and $\hbar$ is Planck's constant. On the basis of studies of the crystal structure[107, 205] and the intermolecular potential[107] for **2** a molecule of **2** is considered to undergo the in-plane 180° flipping, being consistent with the result of the powder pattern. Therefore, C is evaluated only from the intermolecular contributions:

$$C = \frac{3}{640}\gamma^4\hbar^2\sum_{i}\left(\frac{1}{r_{i1}^6} + \frac{1}{r_{i2}^6} + \frac{1 - 3\cos^2\theta_{i12}}{r_{i1}^3 r_{i2}^3}\right), \tag{4-16}$$

where r_{i1} denotes the distance between the protons of the ith intermolecular pair, r_{i2} represents that between the same proton pair after the in-plane 180° flipping of the molecule, and θ_{i12} is the angle between the proton-pair directions in the two states. The three methyl protons are regarded as a condensed nuclei.[108] The temperature dependence of proton T_1 for **1** can be fitted satisfactorily to a single Bloembergen-Purcell-Pound (BPP) equation, eq. (4-4) for $l = 1$, whereas that for **2** is fitted to a combination of two BPP equations [eq. (4-4) for $l = 1, 2$] as shown in Figure 4-8. The least-squares fitted parameters, E_a, τ_0, C, and the calculated values (C^{calc}) for C are listed in Table 4-2. The T_1 plots for **1** were interpreted by assigning the minimum to the methyl rotation. The minima at the higher and the lower temperatures in the T_1 data for **2** were assigned to in-plane 180° flipping and methyl rotation, respectively. These assignments are confirmed by the fact that the calculated and experimental C values agree closely.

To analyze the temperature dependence of T_1 for **3**, one should take into account the in-plane molecular and methyl rotations simultaneously. Because of the low symmetry attendant with the molecular structure of **3**, a rotational diffusion model with the correlation time τ_2 is assumed for the in-plane molecular rotation process.[109,110] Taking into account the methyl rotation with the correlation time τ_1, the overall spin–lattice relaxation for **3** can be written as a sum of three terms:

$$\frac{1}{T_1} = \left(\frac{1}{T_1}\right)_{\text{Me}} + \left(\frac{1}{T_1}\right)_{\text{in-plane}} + \left(\frac{1}{T_1}\right)_{\text{inter}} \tag{4-17}$$

TABLE 4-2. Best-fit Parameters to the T_1 Data for (**1**), (**2**), and (**3**)[a] and the Calculated C (C^{calc}) for **1** and **2**[b]

	E_a (kcal mol^{-1})	τ_0 (s)	C (10^9 s^{-2})	C^{calc} (10^9 s^{-2})	Motion
1	3.22 ± 0.55	$(9.8 \pm 1.7) \times 10^{-14}$	3.76×0.17	3.90	Methyl
2	3.49 ± 0.06	$(1.7 \pm 0.3) \times 10^{-13}$	5.95 ± 0.26	5.84	Methyl
	17.43 ± 1.61	$(1.7 \pm 3.7) \times 10^{-19}$	1.02 ± 0.08	1.00	In-plane
3	4.37 ± 0.12[c]	$(1.7 \pm 0.8) \times 10^{-14}$	0.72 ± 0.08		Methyl
	4.69 ± 0.41[d]	$(5.4 \pm 4.2) \times 10^{-13}$	0.84 ± 0.27		In-plane

[a]Reproduced with Permission from Ref. 50.
[b]Errors are 99% confidence intervals.
[c]Value for C_1.
[d]Value for C_5.

The first term represents the relaxation from the motion of the methyl group:[106,109,110]

$$\left(\frac{1}{T_1}\right)_{\text{Me}} = \frac{N_{\text{Me}}}{N_{\text{all}}}\frac{9}{160}\frac{\gamma^4\hbar^2}{r^6}[3B(\tau_1) + 2B(\tau_2/4) + 4B(\tau_3) + B(\tau_4)] \qquad (4\text{-}18)$$

where:

$$\tau_3^{-1} = \tau_1^{-1} + \tau_2^{-1} \quad \tau_4^{-1} = \tau_1^{-1} + 4\tau_2^{-1} \qquad (4\text{-}19)$$

The second term arises from fluctuation of the dipole–dipole interactions among all protons in a molecule from the in-plane molecular rotation:

$$\left(\frac{1}{T_1}\right)_{\text{in-plane}} = \frac{2}{N_{\text{all}}}\sum_{i>j}\frac{9}{40}\frac{\gamma^4\hbar^2}{r_{ij}^6}B(\tau_2/4) \qquad (4\text{-}20)$$

where the three methyl protons are regarded as condensed nuclei. The third term from the intermolecular contributions is given by:[77,78]

$$\left(\frac{1}{T_1}\right)_{\text{inter}} = C_5 B(\tau_2/8) \qquad (4\text{-}21)$$

where C_5 is the coupling constant for intermolecular dipole–dipole interactions. Therefore, eq. (4-17) is rewritten:

$$\frac{1}{T_1} = C_1 B(\tau_1) + C_2 B(\tau_2/4) + C_3 B(\tau_3) + C_4 B(\tau_4) + C_5 B(\tau_2/8) \qquad (4\text{-}22)$$

Both τ_1 and τ_2 follow eq. (4-6), and the corresponding activation energies, E_a^1 and E_a^2, and pre-exponential factors, τ_0^1 and τ_0^2, are taken to be as adjustable parameters. The ratios of C_1, C_2, C_3, and C_4 is assumed to be 1:1.66:1.33:0.33 on the basis of the geometry of **3**, which is employed for the calculation of the second moment, and is fixed through all temperatures; C_5 is treated as another adjustable parameter. Therefore, T_1 curve for **3** is successfully fitted to eq. (4-22). The best-fit values for the parameters C_1 and C_5 are given in Table 4-2, and the best-fit line is shown as a solid curve in Figure 4-8.

The 1-chloro-8-methylnaphthalene molecule, which has a structure similar to that of **3**, is also found to undergo in-plane molecular rotation ($E_a = 4.51$ kcal mol^{-1}).[50] Hence, it is of much interest that no overall molecular rotation is detected for **1** by the present method. Previously, Lauer et al[111] reported from their NMR study that 1,8-difluoronaphthalene molecules underwent a certain molecular motion in the solid state with an activation energy of 11.7 kcal mol^{-1}. As a result of an x-ray diffraction study[112] this molecular motion is assigned to the strong librational motion perpendicular to the molecular plane. The existence of the in-plane 180° flipping with a large activation energy (16 kcal mol^{-1}) was reported for 1,5-dimethylnaphthalene from measurement of the temperature dependence of ^{1}H spin-lattice relaxation times in the rotating frame.[88] The various types of molecular motion for the substituted naphthalenes described here are related to intermolecular interactions in the crystalline state.

Studies of Static and Dynamic Properties in the Solid State by ^{13}C CPMAS NMR

In recent years, the CPMAS technique has permitted application of NMR techniques not only to insoluble and slightly soluble materials but also to the study of various interesting properties appearing in the solid state. Most of the molecular motions allowed in solution are generally quenched in the solid state, and molecules assume fixed conformations on the balance of various inter- and intramolecular repulsive and attractive interactions; therefore, chemical shifts in the solid state are of significant interest in view of their stereochemical ramifications. This section describes recent ^{13}C CPMAS NMR studies, particularly on the static and dynamic properties of small organic molecules, dividing them into three classes: line shifts and splittings resulting from crystal packing, specific intermolecular interactions, and molecular motions.

Line Shifts and Splittings Resulting from Crystal Packing

Such parameters as chemical shifts and indirect spin–spin coupling constants, which are averaged over many conformations in solution, can be measured as specific values corresponding to fixed molecular conformations.

In compounds with more than two stable crystal structures (polymorphism), the chemical shift of each carbon atom depends on the crystal structure, reflecting the molecular conformation in the crystal. Three different crystal structures (the α, β, and γ forms) of hydroquinone show distinct splitting patterns in their ^{13}C CPMAS NMR spectra.[113] Based on these spectral differences, it can be determined readily whether molecular complexes containing hydroquinone are inclusion compounds or merely stoichiometric complexes. The chemical shift for each carbon atom of (+)-tartaric acid is identical to that for the corresponding carbon atom of (−)-tartaric acid in the racemic compound but is dissimilar to that for the corresponding carbon atom in the pure (+)-tartaric acid crystal because of the difference in their crystal structures.[114] VanderHart[115] has noted that the interior methylene carbons in linear alkanes which take the same, all-trans conformation show the same chemical shift in the three different crystallographic forms, but a slightly downfield shift (1.3 ± 0.4 ppm) in the triclinic form. However, the ^{13}C chemical shifts in the solid state are determined by intramolecular factors to a great extent and can be explained almost in terms of the α, β, and γ effects of the substituents, providing valuable information about conformational properties of molecules.

A singlet line in solution sometimes splits into more than two lines in the solid state. One cause of the splittings is the existence in the unit cell of noncongruent molecules that cannot be superimposed upon each other by any symmetry operation. The existence of noncongruent molecules can be checked by ^{13}C CPMAS NMR spectroscopy[116-118] more readily than by x-ray diffraction. The assignment of the ^{13}C resonance lines to the respective noncongruent molecules is generally difficult. The repeating units in oligomers and such polymers as polysaccharides[119-129] and polypeptides[130-138] may be regarded as noncongruent molecules and show distinct ^{13}C chemical shifts, manifesting conformational differences in the solid state.

Even if only one of the molecular conformations appears in the unit cell, the ^{13}C signal of the carbon atoms that are equivalent in a molecule because of rapid molecular motion in solution often splits by quenching of the dynamic molecular symmetry.[117,118,139-153] This situation is similar to benzaldehydes in solutions at low temperatures,[154-156] where the chemical shifts of the ortho carbons are very different from each other because of coplanar locking of the aldehyde and phenyl groups. Such separations in the solid state were found for the ortho carbons of several hydroxybenzaldehydes:[117] the signal of the ortho carbon proximate to the oxygen atom of the aldehyde group appears upfield by 9–11 ppm relative to that of the other ortho carbon. This large separation is ascribed to the difference between the steric effects of the aldehyde group on the two ortho carbons, providing a useful means of determining conformations in the solid state. Such splitting of the ^{13}C resonance of the ortho carbons has been observed for a number of substituted benzenes,[113,137,138,147] reflecting the differences in the steric interactions between the ortho carbons and the substituents. Conformations of flexible, partially hydrogenated aromatic compounds[118,157-159] have been investigated both in solids and in solutions. The average value of the chemical shifts separated in the solid state is generally different from the value in solution.

Finally the ^{13}C signal for the carbon atom bonded to or near a nitrogen atom splits into an asymmetrical doublet in the ^{13}C CPMAS NMR spectra.[116,129,148] This is because the ^{14}N quadrupole interaction tilts the quantization axis of the ^{14}N spins away from the direction of the applied field so that the ^{13}C–^{14}N dipolar interactions cannot be removed by MAS.[160-165]

Specific Intermolecular Interactions

This section deals with studies of compounds related to intermolecular hydrogen bonding, complexation, inclusion, and adsorption. Various properties of these interactions have not been investigated clearly in solutions owing to dissociation or rapid equilibration.

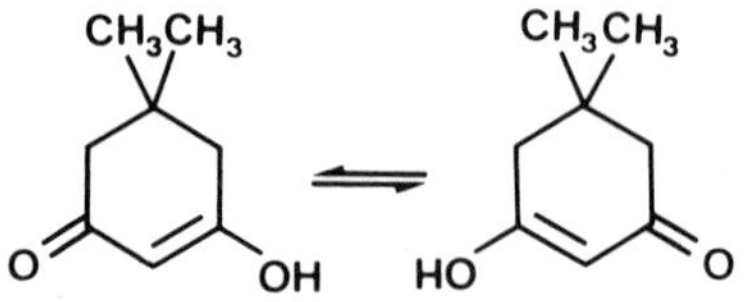

CH_3CH_3 CH_3CH_3
O OH HO O

4

Intermolecular hydrogen bonding results not only in dimeric structures generally observed in carboxylic acid dimers, such as benzoic acid dimer, but also in infinite helical[166,167] or zigzag chain[168] structures, which appear in enol forms of certain 1,3-diketones. Although the carbonyl and enol carbons in the enol form of 5,5-dimethyl-1,3-cyclohexanedione **4** are equivalent in solutions because of the rapid enol–enol tautomerism, their chemical shifts are notably different in the solid state:[169] the chemical shifts of the carbonyl and enol carbons move downfield by 8.1 and 10.2 ppm, respectively, from those for the corresponding carbons of its methyl ether in $(CD_3)_2SO$ solution.[170] The short O···O hydrogen bond distance (2.593–2.595 Å)[166,167] in the solid state is responsible for such large downfield shifts. An inverse correlation between the downfield shifts of the hydrogen-bonded carbonyl carbons and the O···O hydrogen-bond distances in the solid state was found for some hydroxybenzaldehydes.[117] Similar downfield shifts were observed for the carboxylic carbons in some substituted benzoic acids,[147,171] but the correlation is not clear. The ^{13}C line splitting caused by suppression of proton transfer in the solid state was reported for imidazole.[165,172]

In π–π-type charge-transfer complexes containing hexamethylbenzene as a donor molecule, the ^{13}C chemical shifts for the ring carbons of hexamethylbenzene move downfield in proportion to the electron affinities of the acceptor molecules; this suggests the migration of, at most, 10% of electrons in the ground state.[173] The carbon signal of cyclooctatetraene (COT) in pentacarbonylcyclooctatetraenediiron appears as a single line in the ^{13}C CPMAS NMR spectra even at $-160°C$ because of the rapid change of bonding positions among a COT molecule and iron atoms;[174] such a change is a typical feature of the fluxional organometallics.[175] The carbon signal of COT in tetracarbonylbis(cyclooctatetraene)triruthenium appears as a single line at 27°C, whereas at $-180°C$ it splits into a lot of peaks because of freezing of the exchange process.[176] Maricq et al[177] obtained the principal values of the carbon chemical shift tensors for arene chromium tricarbonyl complexes by observing spinning sidebands; the component corresponding to the principal axis along the radial direction of the benzene ring moves distinctly upfield relative to the value without complexation. This arene-type effect contrasts with the fact that in the ring carbon tensors only the principal value corresponding to the principal axis perpendicular to the ring plane is sensitive to the replacement of the substituent[178] (the aryl-type effect).

Conformational changes of host molecules upon inclusion were examined by ^{13}C CPMAS NMR spectra for α- and β-cyclodextrin inclusion complexes[128,129] and β-quinol clathrates.[113,170] Molecular motion of guest molecules containing a nitrile group in several α-cyclodextrin[129] and urea[45] inclusion complexes and acetonitrile-β-quinol clathrate[179] was discussed on the basis of the ^{13}C–^{14}N dipolar splitting. Structural deformation and molecular motion of adsorbed molecules as well as adsorbent molecules were also studied by ^{13}C CPMAS NMR spectroscopy.[180–185]

Molecular Motion

Some types of molecular motions are still allowed in the solid state. However, their amplitudes and frequencies are reduced relative to the corresponding motions in solution because the transition state for molecular motion generally requires wider space than the ground state. Dynamic parameters for molecular motions in the solid state can be obtained using variable-temperature (VT) CPMAS NMR techniques. In evaluating these parameters, however, it should be kept in mind that the path for molecular motion in solids may differ from that in solutions and, moreover, that a degenerate molecular motion in solutions sometimes becomes nondegenerate owing to anisotropic intermolecular interactions in the solid state.

First are described the VT ^{13}C NMR studies of molecular rearrangements that remain rapid in the solid state. Yannoni and co-workers[174,186–188] measured the cryogenic-temperature ^{13}C CP NMR spectra for the 2-norbornyl cation (**5**); C-1 and C-2 are equivalent to each other even at 5 K, whereas C-6 is not equivalent to them.[189] They concluded that the activation energy for the 1,2-hydride shift either between unsymmetrically bridged structures or between unsymmetrically localized structures is less than 0.2 kcal mol^{-1}; the existence of the latter structures has been negated by investigation of the perturbation isotope effects on the ^{1}H chemical shifts.[190] This upper limit is very small in comparison with the value estimated by 50-MHz ^{13}C NMR in solution.[191] Subsequent molecular orbital calculations, including electron correlation,[192,193] indicate that the symmetrically bridged, nonclassical structure is the stable geometry of **5** in the isolated state. Carbons 2 and 3 of the *sec*-butyl cation (**6**) in solid SbF_5 are also equivalent at temperatures as low as $-190°C$ probably because of the fast 2,3-hydride shift of this cation.[194]

Unsymmetrically localized structures

Unsymmetrically bridged structures

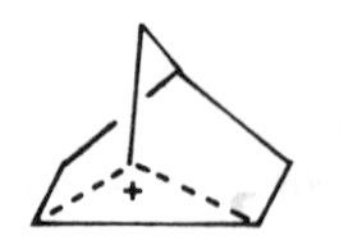

Symmetrically bridged structure

5

$CH_3-\overset{+}{C}H-CH_2-CH_3$

6

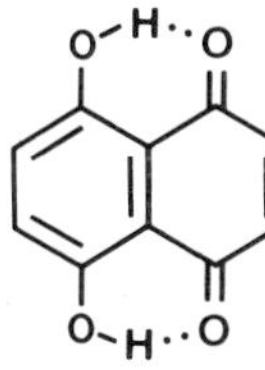

7

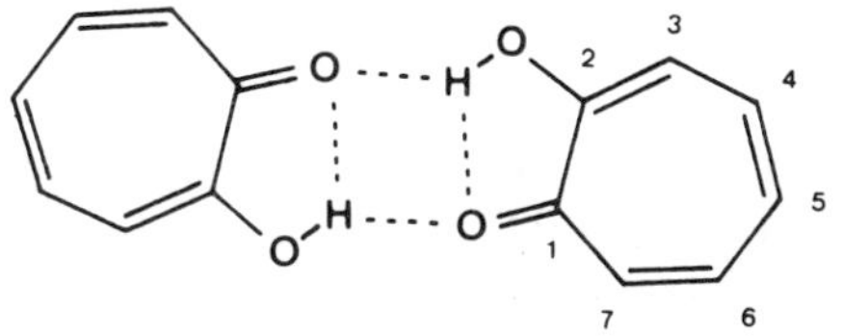

8

Shiau et al[195] have found that the ^{13}C CPMAS NMR spectra of 5,8-dihydroxy-1,4-naphthoquinone (**7**) at intermediate temperatures are the superpositions of the low- and high-temperature limiting spectra in proportions depending on the temperature. They consider that the exchange process is controlled by the occurrence of a second-order phase transition: proton transfer is not allowed in the low-temperature phase, whereas rapid transfer takes place in the high-temperature phase. When proton transfer is accompanied with another molecular motion, the activation energy for the combined process may become very high. Tropolone molecules (**8**) take a bifurcated hydrogen-bonding structure in the solid state.[196] The two-dimensional ^{13}C CPMAS NMR spectrum for tropolone indicates the site exchange between the pair carbons (C-1 and C-2, C-3 and C-7, and C-4 and C-6).[197] This process is explainable by either intra- or intermolecular proton transfer accompanied by an energetically unfavorable out-of-plane molecular rotation. The activation energy for this process, determined by magnetization transfer experiments[46] is notably high: $E_a = 26 \pm 5$ kcal mol^{-1}.

Retardation of the exchange rates in the solid state has been observed for some molecular rearrangements that require relatively large changes of the molecular framework during their processes. The exchange rate for the 6,1,2-hydride shift in **5** was determined by lineshape analysis of ^{13}C CPMAS NMR spectra.[198] It is slower by a factor of three to four in the solid state than in solution. Interestingly, the activation energy (6.1 ± 0.5 kcal mol^{-1}) in the solid state is close to the free energy of activation (5.9 ± 0.2 kcal mol^{-1})[199] in solution. Similarly, the rate for the scrambling of the four carbon atoms of **6** is significantly slower in a solid SbF_5 matrix than in solution.[194] The rate retardation is remarkable for the 1,2-methyl shift in the heptamethylbenzenonium ion (**9**).[200,201] The activation energy for this rearrangement, however, is strikingly low in the solid state, presumably because of electrostatic interaction between the partial positive charge on the migrating methyl group and the neighboring $AlCl_4^-$ anions.[201]

CH₃CH₃, CH₃, CH₃, CH₃, CH₃, CH₃

9 **10**

The degenerate Cope rearrangement of semibulvalene (**10**) in solution[202] becomes nondegenerate in the dynamic phase of the solid state.[203,204] The exchange rates corresponding to the higher barrier are 10–20 times slower than those in solution. However, the free energy of activation for the higher barrier process in the solid state is 6.3 kcal mol^{-1} at -132°C, being comparable to the solution value (5.5 kcal mol^{-1} at -133°C).

Finally, Waugh and co-workers[43,44] have determined the motional parameters for the isotropic overall molecular rotation of adamantane, the in-plane ring rotation of hexamethylbenzene and decamethylferrocene, and the combined $C_3 + C_3'$ motion of hexamethylethane using the line-broadening effects from molecular motions under MAS or rf decoupling.

References

1. Mehring, M. "Principles of High Resolution NMR in Solids"; Springer-Verlag: Berlin, 1983.
2. Hester, R. K.; Ackerman, J. L.; Neff, B. L.; Waugh, J. S. *Phys. Rev. Lett.* **1976**, *36*, 1081.
3. Rybaczewski, E. F.; Neff, B. L.; Waugh, J. S.; Sherfinski, J. S. *J. Chem. Phys.* **1977**, *67*, 1231.
4. Naito, A.; Barker, P. B.; McDowell, C. A. *J. Chem. Phys.* **1984**, *81*, 1583.
5. Aue, W. P.; Bartholdi, E.; Ernst, R. R. *J. Chem. Phys.* **1976**, *64*, 2229.
6. Bax, A. "Two-Dimensional Nuclear Magnetic Resonance in Liquids"; Delft Univ. Press: Delft, 1982.
7. Stoll, M. E.; Vega, A. J.; Vaughan, R. W. *J. Chem. Phys.* **1976**, *65*, 4093.
8. Linder, M.; Höhener, A.; Ernst, R. R. *J. Chem. Phys.* **1980**, *73*, 4959.
9. Munowitz, M. G.; Griffin, R. G.; Bodenhausen, G.; Huang, T. H. *J. Am. Chem. Soc.* **1981**, *103*, 2529.
10. Munowitz, M. G.; Griffin, R. G. *J. Chem. Phys.* **1982**, *76*, 2848.
11. Munowitz, M. G.; Aue, W. P.; Griffin, R. G. *J. Chem. Phys.* **1982**, *77*, 1686.
12. Schaefer, J.; McKay, R. A.; Stejskal, E. O.; Dixon, W. T. *J. Magn. Reson.* **1983**, *52*, 123.
13. Munowitz, M. G.; Huang, T. H.; Dobson, C. M.; Griffin, R. G. *J. Magn. Reson.* **1984**, *57*, 56.
14. Schaefer, J.; Stejskal, E. O.; McKay, R. A.; Dixon, W. T. *J. Magn. Reson.* **1984**, *57*, 85.
15. Terao, T.; Miura, H.; Onodera, T.; Fujii, T.; Saika, A. Proc. 22nd Congr. AMPERE, **1984**, 576.

16. Terao, T.; Fujii, T.; Onodera, T.; Saika, A. *Chem. Phys. Lett.* **1984**, *107*, 145.
17. Yannoni, C. S.; Kendrick, R. D. *J. Chem. Phys.* **1981**, *74*, 747.
18. Horne, D.; Kendrick, R. D.; Yannoni, C. S. *J. Magn. Reson.* **1983**, *52*, 299.
19. Maricq, M. M.; Waugh, J. S. *J. Chem. Phys.* **1979**, *70*, 3300.
20. Herzfeld, J.; Berger, A. E. *J. Chem. Phys.* **1980**, *73*, 6021.
21. Aue, W. P.; Ruben, D. J.; Griffin, R. G. *J. Magn. Reson.* **1981**, *43*, 472.
22. Alla, M. A.; Kundla, E. I.; Lippmaa, E. T. *Pis'ma Zh. Eksp. Teor. Fiz.* **1978**, *27*, 208.
23. Yarif-Agaev, Y.; Tutunjian, P. N.; Waugh, J. S. *J. Magn. Reson.* **1982**, *47*, 51.
24. Bax, A.; Szeverenyi, N. M.; Maciel, G. E. *J. Magn. Reson.* **1983**, *51*, 400.
25. Bax, A.; Szeverenyi, N. M.; Maciel, G. E. *J. Magn. Reson.* **1983**, *52*, 147.
26. Bax, A.; Szeverenyi, N. M.; Maciel, G. E. *J. Magn. Reson.* **1983**, *55*, 494.
27. Terao, T.; Miura, H.; Saika, A. *J. Chem. Phys.* **1981**, *75*, 1573; *J. Magn. Reson.* **1982**, *49*, 365.
28. Terao, T.; Miura, H.; Saika, A. *J. Am. Chem. Soc.* **1982**, *104*, 5228.
29. Zilm, K. W.; Grant, D. M. *J. Magn. Reson.* **1982**, *48*, 524.
30. Mayne, C. L.; Pugmire, R. J.; Grant, D. M. *J. Magn. Reson.* **1984**, *56*, 151.
31. Opella, S. J.; Frey, M. H. *J. Am. Chem. Soc.* **1979**, *101*, 5854.
32. Caravatti, P.; Bodenhausen, G.; Ernst, R. R. *Chem. Phys. Lett.* **1982**, *89*, 363.
33. Caravatti, P.; Braunschweiler, L.; Ernst, R. R. *Chem. Phys. Lett.* **1983**, *100*, 305.
34. Roberts, J. E.; Vega, S.; Griffin, R. G. *J. Am. Chem. Soc.* **1984**, *106*, 2506.
35. Dixon, W. T. *J. Chem. Phys.* **1982**, *77*, 1800.
36. Dixon, W. T.; Schaefer, J.; Sefcik, M. D.; Stejskal, E. O.; McKay, R. A. *J. Magn. Reson.* **1982**, *49*, 341.
37. Torchia, D. A. *J. Magn. Reson.* **1978**, *30*, 613.
38. Schaefer, J.; Stejskal, E. O.; Buchdahl, R. *Macromolecules* **1977**, *10*, 384.
39. VanderHart, D. L.; Garroway, A. N. *J. Chem. Phys.* **1979**, *71*, 2773.
40. Spiess, H. W. In "NMR Basic Principles and Progress", Diehl, P.; Fluck, E.; Kosfeld, R., Eds.; Springer-Verlag: Berlin, 1978: Vol. 15, pp. 55–214.
41. Wemmer, D. E.; Ruben, D. J.; Pines, A. *J. Am. Chem. Soc.* **1981**, *103*, 28.
42. Kuhns, P. L.; Conradi, M. S. *J. Chem. Phys.* **1982**, *77*, 1771.
43. Suwelack, D.; Rothwell, W. P.; Waugh, J. S. *J. Chem. Phys.* **1980**, *73*, 2559.
44. Rothwell, W. P.; Waugh, J. S. *J. Chem. Phys.* **1981**, *74*, 2721.
45. Okazaki, M.; Naito, A.; McDowell, C. A. *Chem. Phys. Lett.* **1983**, *100*, 15.
46. Szeverenyi, N. M.; Bax, A.; Maciel, G. E. *J. Am. Chem. Soc.* **1983**, *105*, 2579.
47. Terao, T.; Matsui, S.; Saika, A. *Chem. Phys. Lett.* **1979**, *64*, 582.
48. Matsui, S.; Terao, T.; Saika, A. *J. Chem. Phys.* **1982**, *77*, 1788.
49. Nagaoka, S.; Terao, T.; Imashiro, F.; Saika, A.; Hirota, N.; Hayashi, S. *Chem. Phys. Lett.* **1981**, *80*, 580.
50. Imashiro, F.; Takegoshi, K.; Okazawa, S.; Furukawa, J.; Terao, T.; Saika, A.; Kawamori, A. *J. Chem. Phys.* **1983**, *78*, 1104.
51. Palin, D. E.; Powell, H. M. *J. Chem. Soc.* **1948**, 571.
52. Hirotsu, K.; Nishimoto, K., unpublished results, 1979.
53. Ripmeester, J. A.; Hawkins, R. E.; Davidson, D. W. *J. Chem. Phys.* **1979**, *71*, 1889.
54. Ditchfield, R. Conference on Critical Evaluation of Chemical and Physical Structural Information, Dartmouth College, NH, June 1973.
55. Pines, A.; Abramson, E. *J. Chem. Phys.* **1974**, *60*, 5130.
56. See, e.g., Das, T. P.; Hahn, E. L. "Nuclear Quadrupole Resonance Spectroscopy, Solid State Physics, Suppl. 1". Academic: New York, 1958.
57. Zweers, A. E.; Brown, H. M. *Physica B* **1977**, *85*, 223.
58. Zweers, A. E.; Brown, H. M.; Huiskamp, W. J. *Physica B*, **1977**, 239.
59. Soda, G. *Kagaku no Ryoiki* **1974**, *28*, 799 (in Japanese).
60. Soda, G.; Chihara, H. Personal communication, 1978.
61. Takeda, S.; Soda, G.; Chihara, H. *Mol. Phys.* **1982**, *47*, 501.
62. Hayashi, S.; Umemura, J. *J. Chem. Phys.* **1974**, *60*, 2630.
63. Hayashi, S.; Kimura, N. *Bull. Inst. Chem. Res. Kyoto Univ.* **1966**, *44*, 335.
64. Sim, G. A.; Robertson, J. M.; Goodwin, T. H. *Acta Crystallogr.* **1955**, *8*, 157.
65. Chang, J. J.; Griffin, R. G.; Pines, A. *J. Chem. Phys.* **1975**, *62*, 4923.
66. Kempf, J.; Spiess, H. W.; Haeberlen, U.; Zimmermann, H. *Chem. Phys. Lett.* **1972**, *17*, 39.
67. van Dongen Torman, J.; Veeman, W. S.; de Boer, E. *J. Magn. Reson.* **1978**, *32*, 49.
68. Scheiner, S.; Kern, C. W. *J. Am. Chem. Soc.* **1979**, *101*, 4081.

69. Bene, J. E. D.; Kochenour, W. L. *J. Am. Chem. Soc.* **1976**, *98*, 2041.
70. Lipinski, J.; Sokalski, W. A. *Chem. Phys. Lett.* **1980**, *76*, 88.
71. Graf, F.; Meyer, R.; Ha, T.-K.; Ernst, R. R. *J. Chem. Phys.* **1981**, *75*, 2914.
72. Nagaoka, S.; Hirota, N.; Matsushita, T.; Nishimoto, K. *Chem. Phys. Lett.* **1982**, *92*, 498.
73. Hayashi, S.; Umemura, J.; Kato, S.; Morokuma, K. *J. Phys. Chem.* **1984**, *88*, 1330.
74. Hayashi, S.; Umemura, J.; Nakamura, R. *J. Mol. Struct.* **1980**, *69*, 123.
75. Meyer, B. H.; Graf, F.; Ernst, R. R. *J. Chem. Phys.* **1982**, *76*, 767.
76. Nagaoka, S.; Terao, T.; Imashiro, F.; Saika, A.; Hirota, N.; Hayashi, S. *J. Chem. Phys.* **1983**, *79*, 4694.
77. Anderson, J. E. *J. Chem. Phys.* **1965**, *43*, 3575.
78. Haeberlen, U.; Maier, G. *Z. Naturforsch.* **1967**, *22A*, 1236.
79. Andrew, E. R.; Eades, R. G. *Proc. R. Soc. London Ser. A.* **1953**, *218*, 537.
80. Eveno, M.; Meinnel, J. *J. Chim. Phys. Phys.-Chim. Biol.* **1966**, *63*, 108.
81. Anderson, J. E.; Slichter, W. P. *J. Chem. Phys.* **1966**, *44*, 1797.
82. Allen, P. S.; Cowking, A. *J. Chem. Phys.* **1967**, *47*, 4286.
83. Jones, G. P.; Eades, R. G.; Terry, K. W.; Llewellyn, J. P. *J. Phys. C* **1968**, *1*, 415.
84. Bernard, H. W.; Tanner, J. E.; Aston, J. G. *J. Chem. Phys.* **1969**, *50*, 5016.
85. Albert, S.; Gutowsky, H. S.; Ripmeester, J. A. *J. Chem. Phys.* **1972**, *56*, 2844.
86. Boden, N.; Gibb, M. *Mol. Phys.* **1974**, *27*, 1359.
87. Albert, S.; Ripmeester, J. A. *J. Chem. Phys.* **1979**, *70*, 1352.
88. von Schütz, J. U.; Weithause, M. *Z. Naturforsch.* **1975**, *30A*, 1302.
89. Bondi, A. *J. Phys. Chem.* **1964**, *68*, 441.
90. Fyfe, C. A.; Gilson, D. F. R.; Thompson, K. H. *Chem. Phys. Lett.* **1970**, *5*, 215.
91. Fyfe, C. A.; Kupferschmidt, G. J. *Can. J. Chem.* **1973**, *51*, 3774.
92. Fyfe, C. A.; Dunnell, B. A.; Ripmeester, J. *Can. J. Chem.* **1971**, *49*, 3332.
93. von Schütz, J. U.; Wolf, H. C. *Z. Naturforsch.* **1972**, *27A*, 42.
94. Rushworth, F. A. *J. Chem. Phys.* **1952**, *20*, 920.
95. Veeman, W. S. *Phil. Trans. R. Soc. London* **1981**, *A299*, 629.
96. Pines, A.; Gibby, M. G.; Waugh, J. S. *Chem. Phys. Lett.* **1972**, *15*, 373.
97. Pausak, S.; Pines, A.; Waugh, J. S. *J. Chem. Phys.* **1973**, *59*, 591.
98. Pausak, S.; Tegenfeld, J.; Waugh, J. S. *J. Chem. Phys.* **1974**, *61*, 1021.
99. van Dongen Torman, J.; Veeman, W. S. *J. Chem. Phys.* **1978**, *68*, 3233.
100. Linder, M.; Höhener, A.; Ernst, R. R. *J. Magn. Reson.* **1979**, *35*, 379.
101. Strub, H.; Beeler, A. J.; Grant, D. M.; Michl, J.; Cutts, P. W.; Zilm, K. W. *J. Am. Chem. Soc.* **1983**, *105*, 3333.
102. Bright, D.; Maxwell, I. E.; de Boer, J. *J. Chem. Soc., Perkin Trans. 2* **1973**, 2101.
103. Stejskal, E. O.; Gutowsky, H. S. *J. Chem. Phys.* **1958**, *28*, 388.
104. Anderson, J. E.; Slichter, W. P. *J. Phys. Chem.* **1965**, *69*, 3099.
105. Stohrer, M.; Noack, F. *J. Chem. Phys.* **1977**, *67*, 3729.
106. Woessner, D. E. *J. Chem. Phys.* **1962**, *36*, 1.
107. Imashiro, F; Takegoshi, K.; Saika, A.; Taira, Z.; Asahi, Y. *J. Am. Chem. Soc.* **1985**, *107*, 2341.
108. Albert, S.; Gutowsky, H. S.; Ripmeester, J. A. *J. Chem. Phys.* **1972**, *56*, 3672.
109. Dunn, M. B.; McDowell, C. A. *Mol. Phys.* **1972**, *24*, 969.
110. Bloembergen, N. *Phys. Rev.* **1956**, *104*, 1542.
111. Lauer, O.; Stehlik, D.; Hausser, K. H. *J. Magn. Reson.* **1972**, *6*, 524.
112. Meresse, A.; Courseille, C.; Leroy, F.; Chanh, N. B. *Acta Crystallogr.* **1975**, *B31*, 1236.
113. Ripmeester, J. A. *Chem. Phys. Lett.* **1980**, *74*, 536.
114. Hill, H. D. W.; Zens, A. P.; Jacobus, J. *J. Am. Chem. Soc.* **1979**, *101*, 7090.
115. VanderHart, D. L. *J. Magn. Reson.* **1981**, *44*, 117.
116. Balimann, G. E.; Groombridge, C. J.; Harris, R. K.; Packer, K. J.; Say, B. J.; Tanner, S. F. *Phil. Trans. R. Soc. London* **1981**, *A299*, 643.
117. Imashiro, F.; Maeda, S.; Takegoshi, K.; Terao, T.; Saika, A. *Chem. Phys. Lett.* **1983**, *99*, 189.
118. Ariel, S.; Scheffer, J. R.; Trotter, J.; Wong, Y.-F. *Tetrahedron Lett.* **1983**, *24*, 4555.
119. Schaefer, J.; Stejskal, E. O. *J. Am. Chem. Soc.* **1976**, *98*, 1031.
120. Atalla, R. H.; Gast, J. C.; Sindorf, D. W.; Bartuska, V. J.; Maciel, G. E. *J. Am. Chem. Soc.* **1980**, *102*, 3249.
121. Earl, W. L.; VanderHart, D. L. *J. Am. Chem. Soc.* **1980**, *102*, 3251.
122. Saitô, H.; Tabeta, R.; Harada, T. *Chem. Lett.* **1981**, 571.
123. Saitô, H.; Tabeta, R. *Chem. Lett.* **1981**, 713.

124. Saitô, H.; Tabeta, R.; Hirano, T. *Chem. Lett.* **1981**, 1479.
125. Earl, W. L.; VanderHart, D. L. *J. Magn. Reson.* **1982**, *48*, 33.
126. Jeffrey, G. A.; Wood, R. A.; Pfeffer, P. E.; Hicks, K. B. *J. Am. Chem. Soc.* **1983**, *105*, 2128.
127. Dudley, R. L.; Fyfe, C. A.; Stephenson, P. J.; Deslandes, Y.; Hamer, G. K.; Marchessault, R. H. *J. Am. Chem. Soc.* **1983**, *105*, 2469.
128. Saitô, H.; Izumi, G.; Mamizuka, T.; Suzuki, S.; Tabeta, R. *J. Chem. Soc., Chem. Commun.* **1982**, 1386.
129. Okazaki, M.; McDowell, C. A. *Chem. Phys. Lett.* **1983**, *102*, 20.
130. Opella, M. H.; Frey, M. H.; Cross, T. A. *J. Am. Chem. Soc.* **1979**, *101*, 5856.
131. Pease, L. G.; Frey, M. H.; Opella, S. J. *J. Am. Chem. Soc.* **1981**, *103*, 467.
132. Saitô, H.; Iwanaga, Y.; Tabeta, R.; Narita, M.; Asakura, T. *Chem. Lett.* **1983**, 427.
133. Saitô, H.; Tabeta, R.; Ando, I.; Ozaki, T.; Shoji, A. *Chem. Lett.* **1983**, 1437.
134. Flippen-Anderson, J. L.; Gilardi, R.; Karle, I. L.; Frey, M. H.; Opella, S. J.; Gierasch, L. M.; Goodman, M.; Madison, V.; Dalaney, N. G. *J. Am. Chem. Soc.* **1983**, *105*, 6609.
135. Taki, T.; Yamashita, S.; Satoh, M.; Shibata, A.; Yamashita, T.; Tabeta, R.; Saitô, H. *Chem. Lett.* **1981**, 1803.
136. Kessler, H.; Bermel, W.; Förster, H. *Angew. Chem. Int. Ed. Engl.* **1982**, *21*, 689.
137. Lippmaa, E. T.; Alla, M. A.; Pehk, T. J.; Engelhardt, G. *J. Am. Chem. Soc.* **1978**, *100*, 1929.
138. Chippendale, A. M.; Mathias, A.; Harris, R. K.; Packer, K. J.; Say, B. J.; *J. Chem. Soc., Perkin Trans. 2* **1981**, 1031.
139. Steger, T. R.; Stejskal, E. O.; McKay, R. A.; Stults, B. R.; Schaefer, J. *Tetrahedron Lett.* **1979**, 295.
140. Iverson, D. J.; Hunter, G.; Blount, J. F.; Damewood, J. R. Jr.; Mislow, K. *J. Am. Chem. Soc.* **1981**, *103*, 6073.
141. Bax, A.; Szeverenyi, N. M.; Maciel, G. E. *J. Magn. Reson.* **1982**, *50*, 227.
142. Kawada, Y.; Iwamura, H. *J. Am. Chem. Soc.* **1983**, *105*, 1449.
143. Alemany, L. B.; Grant, D. M.; Pugmire, R. J.; Alger, T. D.; Zilm, K. W. *J. Am. Chem. Soc.* **1983**, *105*, 2133.
144. Alemany, L. B.; Grant, D. M.; Pugmire, R. J.; Alger, T. D.; Zilm, K. W. *J. Am. Chem. Soc.* **1983**, *105*, 2142.
145. Chippendale, A. M.; Mathias, A.; Aujla, R. S.; Harris, R. K.; Packer, K. J.; Say, B. J. *J. Chem. Soc., Perkin Trans. 2* **1983**, 1357.
146. Takegoshi, K.; Imashiro, F.; Terao, T.; Saika, A. *J. Chem. Phys.* **1984**, *80*, 1089.
147. Hays, G. R. *J. Chem. Soc., Perkin Trans. 2* **1983**, 1049.
148. Frey, M. H.; Opella, S. J. *J. Chem. Soc., Chem. Commun.* **1980**, 474.
149. Maciel, G. E.; Shatlock, M. P.; Houtchens, R. A.; Caughey, W. S. *J. Am. Chem. Soc.* **1980**, *102*, 6884.
150. Hays, G. R.; Huis, R.; Coleman, B.; Clague, D.; Verhoeven, J. W.; Rob, F. *J. Am. Chem. Soc.* **1981**, *103*, 5140.
151. Brown, C. E. *J. Am. Chem. Soc.* **1982**, *104*, 5608.
152. Brown, C. E.; Roerig, S. C.; Fujimoto, J. M.; Burger, V. T. *J. Chem. Soc., Chem. Commun.* **1983**, 1506.
153. Hexem, J. G.; Frey, M. H.; Opella, S. J. *J. Chem. Soc., Chem. Commun.* **1983**, *105*, 5718.
154. Lunazzi, L.; Macciantelli, D.; Boicelli, A. C. *Tetrahedron Lett.* **1975**, 1205.
155. Drakenberg, T.; Jost, R.; Sommer, J. M. *J. Chem. Soc., Perkin Trans. 2* **1975**, 1682.
156. Drakenberg, T.; Sommer, J.; Jost, R. *J. Chem. Soc., Perkin Trans. 2* **1980**, 363.
157. Dalling, D. K.; Zilm, K. W.; Grant, D. M.; Heeschen, W. A.; Horton, W. J.; Pugmire, R. J. *J. Am. Chem. Soc.* **1981**, *103*, 4817.
158. McDowell, C. A.; Naito, A.; Scheffer, J. R.; Wong, Y.-F. *Tetrahedron Lett.* **1981**, *22*, 4779.
159. Morin, F. G.; Horton, W. J.; Grant, D. M.; Dalling, D. K.; Pugmire, R. J. *J. Am. Chem. Soc.* **1983**, *105*, 3992.
160. Lippmaa, E. T.; Alla, M.; Roude, H.; Teeaar, R.; Heinmaa, I.; Kundla, E. "Magnetic Resonance and Related Phenomena", *Proc. Congr. AMPERE*, 20th, **1979**, p. 87.
161. Kundla, E.; Alla, M. "Magnetic Resonance and Related Phenomena", *Proc. Congr. AMPERE*, 20th, **1979**; p. 92.
162. Hexem, J. G.; Frey, M. H.; Opella, S. J. *J. Am. Chem. Soc.* **1981**, *103*, 224.
163. Naito, A.; Ganapathy, S.; McDowell, C. A. *J. Chem. Phys.* **1981**, *74*, 5393.
164. Hexem, J. G.; Frey, M. H.; Opella, S. J. *J. Chem. Phys.* **1982**, *77*, 3847.
165. Naito, A.; Ganapathy, S.; McDowell, C. A. *J. Magn. Reson.* **1982**, *48*, 367.

166. Semmingsen, D. *Acta Chem. Scand.* **1974**, *B28*, 169.
167. Singh, I.; Calvo, C. *Can. J. Chem.* **1975**, *53*, 1046.
168. Semmingsen, D. *Acta Chem. Scand.* **1977**, *B31*, 114.
169. Imashiro, F.; Maeda, S.; Takegoshi, K.; Terao, T.; Saika, A. *Chem. Phys. Lett.* **1982**, *92*, 642.
170. Imashiro, F.; Maeda, S.; Takegoshi, K.; Terao, T.; Saika, A. To be published.
171. Imashiro, F.; Maeda, S.; Takegoshi, K.; Terao, T.; Saika, A. Unpublished results (1982).
172. Ganapathy, S.; Naito, A.; McDowell, C. A. *J. Am. Chem. Soc.* **1981**, *103*, 6011.
173. Blann, W. G.; Fyfe, C. A.; Lyerla, J. R.; Yannoni, C. S. *J. Am. Chem. Soc.* **1981**, *103*, 4030.
174. Lyerla, J. R.; Yannoni, C. S.; Fyfe, C. A. *Acc. Chem. Res.* **1982**, *15*, 208.
175. Cotton, F. A. *Acc. Chem. Res.* **1968**, *1*, 257.
176. Lyerla, J. R.; Fyfe, C. A.; Yannoni, C. S. *J. Am. Chem. Soc.* **1979**, *101*, 1351.
177. Maricq, M. M.; Waugh, J. S.; Fletcher, J. L.; McGlinchey, M. J. *J. Am. Chem. Soc.* **1978**, *100*, 6902.
178. Höhener, A. *Chem. Phys. Lett.* **1978**, *53*, 97.
179. Ripmmester, J. A.; Tse, J. S.; Davidson, D. W. *Chem. Phys. Lett.* **1982**, *86*, 428.
180. Dawson, W. H.; Kaiser, S. W.; Ellis, P. D.; Inners, R. R. *J. Am. Chem. Soc.* **1981**, *103*, 6780.
181. Dawson, W. H.; Kaiser, S. W.; Ellis, P. D.; Inners, R. R. *J. Phys. Chem.* **1982**, *86*, 867.
182. Boxhoorn, G.; van Sanfen, R. A.; van Erp, W. A.; Hays, G. R.; Huis, R.; Clague, D. *J. Chem. Soc., Chem. Commun.* **1982**, 264.
183. Sindorf, D. W.; Maciel, G. E. *J. Am. Chem. Soc.* **1983**, *105*, 1848.
184. Sindorf, D. W.; Maciel, G. E. *J. Am. Chem. Soc.* **1983**, *105*, 3767.
185. Maciel, G. E.; Haw, J. F.; Chuang, I.-S.; Hawkins, B. L.; Early, T. A.; McKay, D. R.; Petrakis, L. *J. Am. Chem. Soc.* **1983**, *105*, 5529.
186. Fyfe, C. A.; Lyerla, J. R.; Yannoni, C. S. *J. Am. Chem. Soc.* **1978**, *100*, 5635.
187. Fyfe, C. A.; Mossbrugger, H.; Yannoni, C. S. *J. Magn. Reson.* **1979**, *36*, 61.
188. Macho, V.; Kendrick, R.; Yannoni, C. S. *J. Magn. Reson.* **1983**, *52*, 450.
189. Yannoni, C. S.; Macho, V.; Myhre, P. C. *J. Am. Chem. Soc.* **1982**, *104*, 7380.
190. Saunders, M.; Kates, M. R. *J. Am. Chem. Soc.* **1983**, *105*, 3571.
191. Olah, G. A.; Prakash, G. K. S.; Arranaghi, M.; Anet, F. A. L. *J. Am. Chem. Soc.* **1982**, *104*, 7105.
192. Raghavachari, K.; Haddon, R. C.; Schleyer, P. v. R.; Schaefer, H. F. III *J. Am. Chem. Soc.* **1983**, *105*, 5915.
193. Yoshimine, M.; McLean, A. D.; Liu, B.; DeFrees, D. J.; Binkley, J. S. *J. Am. Chem. Soc.* **1983**, *105*, 6185.
194. Myhre, P. C.; Yannoni, C. S. *J. Am. Chem. Soc.* **1981**, *103*, 230.
195. Shiau, W.-I.; Duesler, E. N.; Paul, I. C.; Curtin, D. Y.; Blann, W. G.; Fyfe, C. A. *J. Am. Chem. Soc.* **1980**, *102*, 4546.
196. Shimanouchi, H.; Sasada, Y. *Acta Crystallogr.* **1973**, *B29*, 81.
197. Szeverenyi, N. M.; Sullivan, M. J.; Maciel, G. E.; *J. Magn. Reson.* **1982**, *47*, 462.
198. Yannoni, C. S.; Macho, V.; Myhre, P. C. *J. Am. Chem. Soc.* **1982**, *104*, 907.
199. Olah, G. A.; White, A. M.; DeMember, J. R.; Commeyras, A.; Lui, C. Y. *J. Am. Chem. Soc.* **1970**, *92*, 4627.
200. Lyerla, J. R.; Yannoni, C. S.; Bruck, D.; Fyfe, C. A. *J. Am. Chem. Soc.* **1979**, *101*, 4770.
201. Borodkin, G. I.; Nagy, S. M.; Mamatyuk, V. I.; Shakirov, M. M.; Shubin, V. G. *J. Chem. Soc., Chem. Commun.* **1983**, 1533.
202. Cheng, A. K.; Anet, F. A. L.; Mioduski, J.; Meinwald, J. *J. Am. Chem. Soc.* **1974**, *96*, 2887.
203. Miller, R. D.; Yannoni, C. S. *J. Am. Chem. Soc.* **1980**, *102*, 7396.
204. Macho, V.; Miller, R. D.; Yannoni, C. S. *J. Am. Chem. Soc.* **1983**, *105*, 3735.
205. Shiner, S.; Noordik, J.; Fisher, A. M.; Eckley, D. M.; Bodenhamer, J.; Haltiwanger, R. C. *Acta Crystallogr.* **1984**, *C40*, 540.

5

TWO-DIMENSIONAL NMR TECHNIQUES TO DETERMINE MOLECULAR SKELETON AND PARTIAL STRUCTURES OF ORGANIC SUBSTANCES

Kuniaki Nagayama

BIOMETROLOGY LAB, JEOL LTD.
NAKAGAMI, AKISHIMA
TOKYO 196, JAPAN

Introduction

In such a rapidly developing field as two-dimensional (2D) NMR spectroscopy, it is not easy to write a comprehensive review covering theoretical aspects to practical applications. To avoid superficiality, this review has been written in accordance with the following outline:

1. Using basic 2D NMR, called 2D shift correlation spectroscopy (COSY), the essential jump from 1D to 2D NMR is explained.
2. Next, the evolution of various ideas and developments of 2D techniques are described.
3. Third, one capability of the technique to elucidate partial structures of organic molecules is discussed. The highly selective information available by analysis of 2D spectral patterns is emphasized.
4. Fourth, various 2D techniques used to connect the backbone skeleton of a

NMR in Stereochemical Analysis

molecule are compared. The problem of mutual connection of partial structures through quaternary carbons is discussed critically.

5. Finally, a difficulty arising from T_2 relaxation is discussed. The future use of 2D NMR in stereochemistry is overviewed.
6. A brief survey of recent applications of 2D NMR in biological systems appears in the Appendix.

Each of the sections is ordered according to the above outline. It is hoped that this small review can afford useful knowledge to many NMR spectroscopists who are not very familiar with this practical and powerful technique.

Basic 2D Technique

Although there has been some debate about who was the founder of 2D NMR, it can be said that this fertile field was initiated with the appearance of a historic paper published by Ernst and his co-workers in 1976.[1] Many farsighted ideas presented in that article became the basis of the rapid development of the technique that ensued. Two new experimental concepts compared with the conventional NMR technique were described in the paper, ie, (1) a rigorous experimental formalism to treat the response of interacting spin systems after multipulse excitation, and (2) introduction of the 2D frequency spectrum to visualize clearly and easily the complicated NMR response. With this treatment the apparently pathologic behavior of spin dynamics in the interacting spin systems can be understood rationally. The basic 2D concept will be explained in two steps corresponding to the two new experimental concepts.

Let us first consider how to generate a 2D spectrum from a one-dimensional (1D) NMR signal. Although various ways can be considered, the most straightforward extension of a 1D spectrum to a 2D spectrum is creation of a mountain-like peak on the frequency maps as shown in Figure 5-1. We are already familiar with the 1D spectrum obtained by Fourier transformation (FT) from the free induction decay (FID) signal in the time domain. What then are the time-domain data corresponding to the 2D peak? These are obtained by time shifting the original FID systematically and aligning the shifted FIDs in the direction of the second time axis with the coordinate of the corresponding time-shift amount. The appearance of 2D FID thus obtained is shown in Figure 5-1 as denoted by the phase modulation. With different t_1, FID changes its initial phase from 0 to $\omega_0 t_1$. The FID signal becomes a mixture of two trigonometric functions. If we divide the phase-shifted FID into cosine and sine components and leave only the cosine components, then we have another 2D FID of amplitude modulation (Figure 5-1). Both types of modulations introduce a 2D peak that has the same resonance frequencies (ω_0) in both frequency axes after 2D FT (2D FT can be done as the first FT in one time axis and then the second FT in another time axis). Without aid of the mathematical

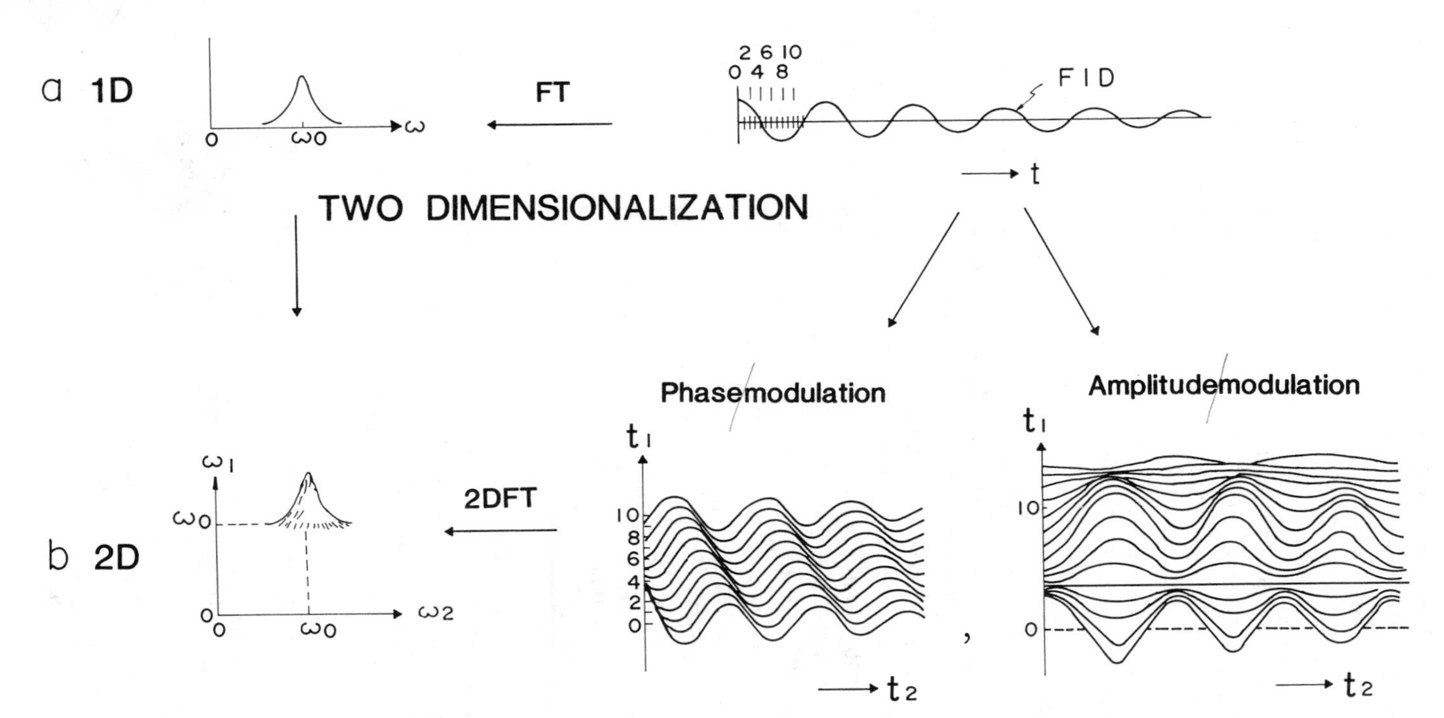

Figure 5-1. A 2D spectrum as an extension of a 1D spectrum. Two possible extensions from a 1D signal in the time domain are shown.

decomposition, what must next be considered is how to get amplitude modulation via a purely experimental procedure.

As already explained, time shifting of NMR data is the first step of 2D extension. This process is a kind of time sharing of time events, which automatically creates two time variables and necessarily introduces two corresponding frequency variables (Figure 5-2a). The 2D FID thus obtained brings about a 2D peak that is characterized by a lineshape shown in Figure 5-2b; ie, absorption and dispersion lineshapes are mixed in both dimensions. To avoid this complication and also to avoid long tailing of 2D peaks, the pure absorption signal should be recorded separately. This can be done by feeding the second radiofrequency (rf) pulse with 90° flip angle before starting detection (Figure 5-2c). By application of this pulse, the magnetization component corresponding to the dispersion mode turns to the direction of Zeeman field (z direction) and contributes nothing to the obtained FID. In the t_2 period, FID always shows pure phase of cosine (or sine depending on the phase detector setting) leading the pure absorption signal shown in Figure 5-2c. This type of 2D FID is nothing other than the amplitude modulation shown in Figure 5-1b.

Trouble, however, occurs when this two-pulse excitation method is applied to a measured system that consists of interacting spins that appear in the NMR spectra as fine structures. As mentioned in the preceding section, complicated NMR responses (spectra) should be observed for interacting spin systems. However, this complication can be overcome and simplified by 2D spectral representation, thereby affording useful information about the NMR parameters and topologies of interacted spins.[1] In the basic 2D NMR, new information arises via crossing or mixing of resonance frequencies among mutually interacting spins. In Figure 5-2d this situation is shown schematically, where different resonance frequencies of two interacting spins characterize one 2D peak (crosspeak).

In Figure 5-3 a simple picture is presented to explain why the crosspeak appears from the two-pulse excitation in the interacting spin system. The most essential point of 2D NMR, magnetization transfer from one spin to the interacting counterspin, is classically interpreted with the combined use of the energy level diagram and the magnetization vector diagram; the picture itself is self-explanatory. The essence of 2D NMR is concentrated in the process

---

Figure 5-2. Two-dimensionalization of 1D signals and corresponding 2D frequency spectra. (a) Two dimensionalization of time by time sharing. (b) Two dimensionalization of a 1D signal by simple time sharing. This method brings about phase modulation of the 2D spectrum where the lineshapes of 2D peaks are mixed by absorption and dispersion. (c) Two dimensionalization of a 1D signal by time sharing and the synchronized second pulse. This method converts phase modulation to amplitude modulation; the pure absorption lineshape results. (d) A characteristic 2D response of a two-pulse excitation (method c) in an AX spin system. A corresponding crosspeak results in the 2D spectrum with pure phase (absorption or dispersion). (e) Schematic of how the second 90° pulse converts phase modulation to amplitude modulation for the 2D response.

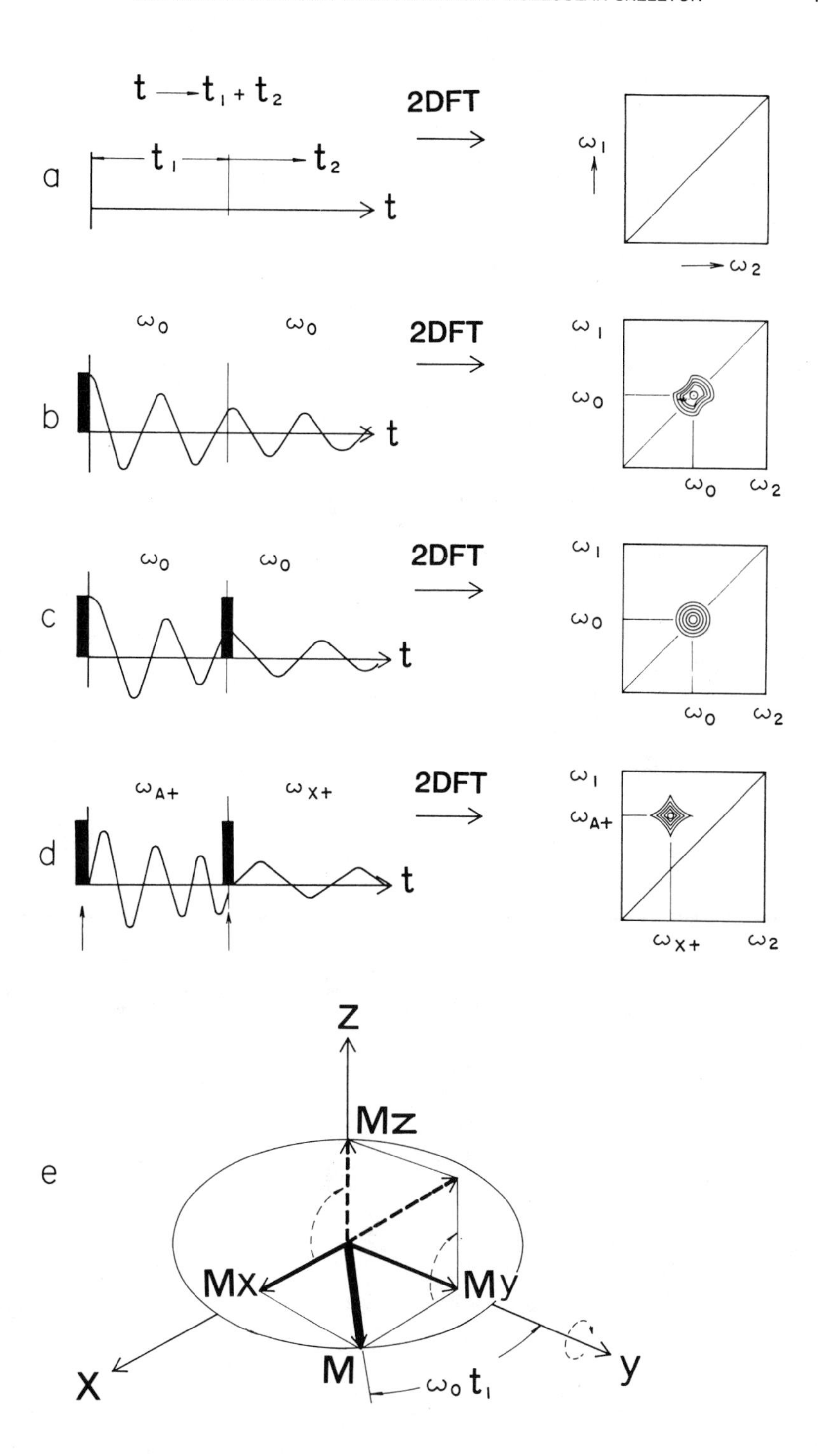

t → t₁ + t₂
a
t₁
t₂
t
2DFT
ω₁
ω₂
b
ω₀
ω₀
t
2DFT
ω₁
ω₀
ω₀
ω₂
c
ω₀
ω₀
t
2DFT
ω₁
ω₀
ω₀
ω₂
d
ωA+
ωX+
t
2DFT
ω₁
ωA+
ωX+
ω₂
e
Z
Mz
Mx
My
M
X
y
ω₀t₁

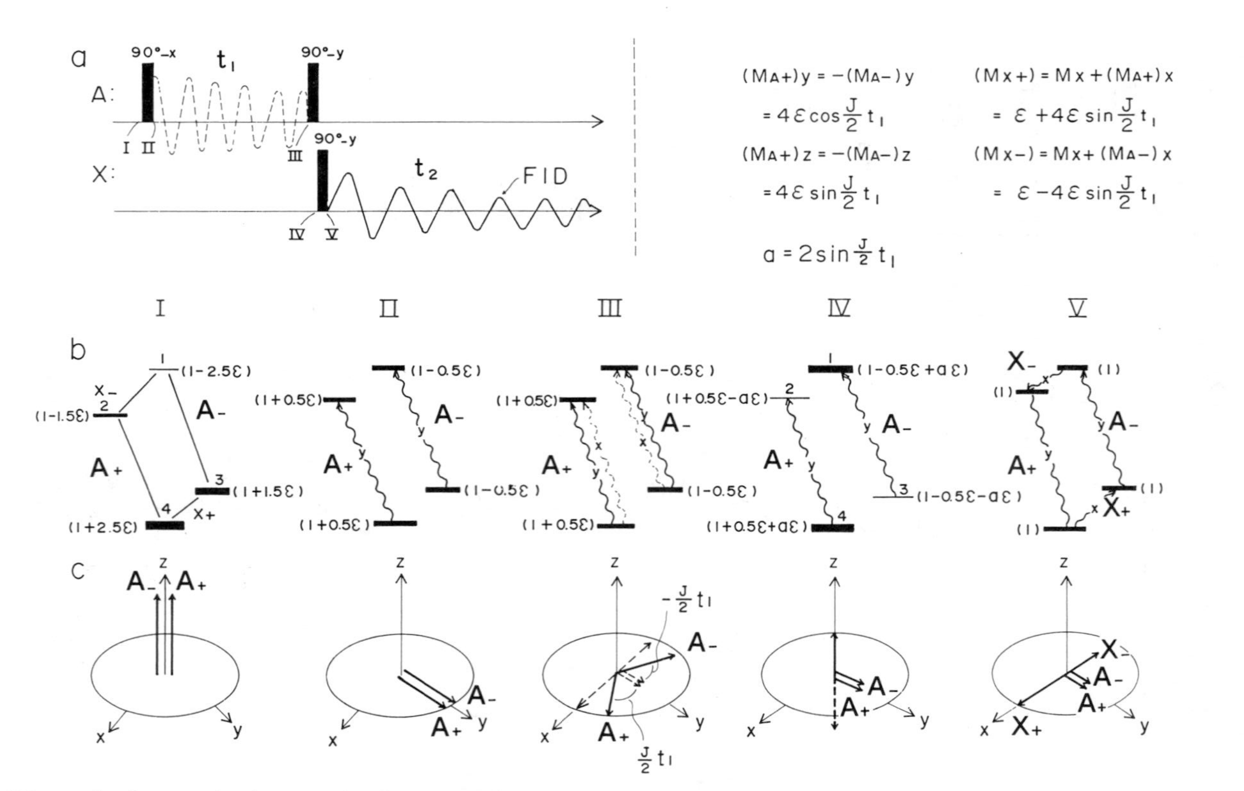

Figure 5-3. Schematic of magnetization transfer that creates 2D crosspeaks. (a) Pulse sequence for the magnetization transfer from *A* to *X* spin. (b) Populations and transitions at the experimental steps I-V [see (a)] represented in an energy level diagram. (c) Magnetization vectors at steps I-V. Note that *A* and *X* magnetizations are drawn pairwise for the + components, respectively.

visualized in the figure from 5-3III to 5-3V, where the two x components of A magnetizations ($A_{\pm}$) modulate the initial amplitudes of the other two x components of X magnetizations ($X_{\pm}$). Readers who are interested in obtaining a more sophisticated physical picture for this process can refer to a recently developed conceptual device, "product operator formalism".[2] Without losing physical rigidity we can trace out intuitively all of the important 2D features.

Derivative 2D Techniques

The very first 2D experiments separated the chemical shifts and the J-coupling (J is the coupling constant) of interacting spins onto the different frequency axes of the 2D spectral domain. The method called homo (or hetero) 2D J-resolved spectroscopy[3–5] (2DJ) is now known to be derived from the basic 2D NMR by simple modification of the original 2D experiment. In such a manner almost all the essential developments in 2D NMR were carefully predicted in the basic 2D design.[1] Therefore, many 2D ideas proposed after the basic 2D NMR (COSY) should be related logically to the basic 2D idea. Here, the various transformation rules implicated in the experimental procedures and their relationship to the basic idea are made clear.

First 2DJ is derived from COSY. When the flip angle of the second pulse in COSY (Figure 5-2d) is changed from 90° to the other angle, how are the 2D peaks differentiated? Consider the case of a 2D spectrum of AX spin (Figure 5-4a). When the flip angle is decreased from 90°, 2D peaks, others than the four autopeaks on the diagonal line decrease their intensities, and only the diagonal peaks remain when finally the second pulse vanishes (flip angle = 0°), as explained in Figure 5-2b. In contrast 2D peaks, including diagonal peaks other than the four autopeaks off the diagonal line, decrease their intensities when the flip angle is increased toward 180°. At the angle of 180° only the off-diagonal autopeaks remain; this result implies that transfer of magnetization occurs between the split spin components, say A_{+} to A_{-}, with 100% efficiency (Figure 5-4b).

This modification of COSY, however, does not complete the derivation of 2DJ. The unnecessary spectral region far from the diagonal line (shaded region in Figure 5-4b), where 2D peaks no longer manifest themselves, must be discarded. This can be accomplished by delaying the start of data acquisition in the t_2 period until after the delay of the same t_1 duration. This delayed acquisition brings about similarity transformation (ST) to the 2D spectrum as shown in Figure 5-4c.[6] The 2D spectrum that develops in the direction of the second frequency axis (ω_1), with relatively small frequencies of J couplings, is further manipulated mathematically with tilting (Figure 5-4d); this is another example of ST. With these three steps the 2DJ spectrum is finally obtained from COSY.

Leaving the flip angle of the second pulse at 90° and employing delayed

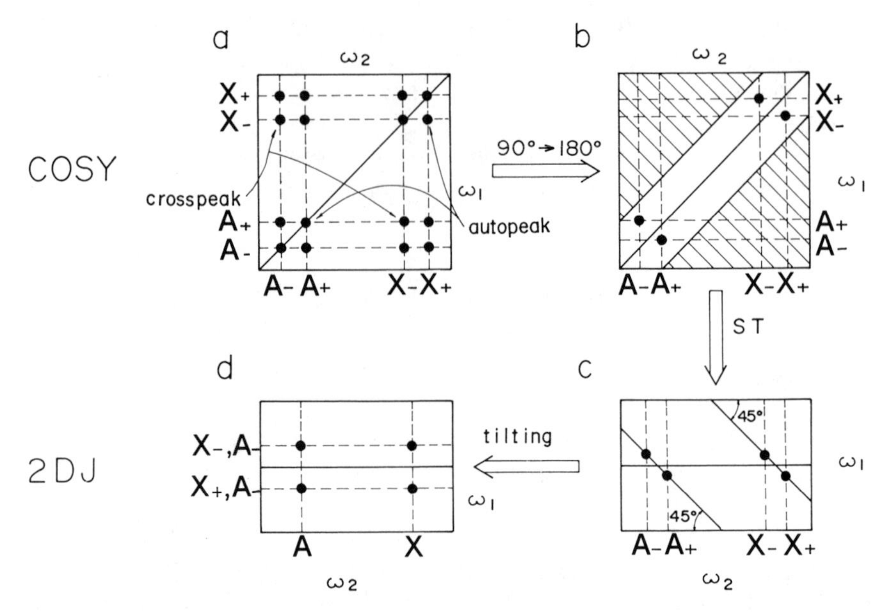

Figure 5-4. Transformation of experimental schemes from COSY to 2DJ. (a) The COSY for an *AX* spin system. (b) Spectral change brought about by alteration of the flip angle of the second pulse from 90° to 180°. (c) Transformation (similarity transformation, ST) of the 2D spectrum associating with delayed data acquisition (spin-echo method). (d) Tilting (ST in the frequency domain) to make the directions of chemical shifts and the *J* splittings perpendicular.

acquisition, we obtain the spin echo correlated spectroscopy (SECSY) spectrum.[6] In SECSY, nothing is changed except the start of the data acquisition. Therefore, COSY and SECSY should be totally equivalent in informational content. However, one big difference between COSY and SECSY should be mentioned, ie, the lineshape of the 2D peaks. To distinguish between positive and negative frequencies in the ω_1 direction, a proper phase alteration of the second rf pulse normally is employed in SECSY.[6] This phase cycling brings about alteration of COSY response from amplitude modulation to phase modulation type. The COSY 2D peaks have a pure absorption lineshape, but SECSY 2D peaks display a mixed-phase lineshape. This is thought to be a drawback, if there is no interest in distinguishing between the signs of ω_1 frequencies. Recently, with use of a purging pulse technique that throws out the impairing dispersion component in 2D peaks, the quality of 2DJ and of SECSY spectra is improved.[7]

In extending the homonuclear 2D technique to the heteronuclear 2D case, all of the original 2D concepts can be applied. Moreover, new experimental techniques characteristic to heterospins simplify the spectral appearance. For the case of ^{1}H–^{13}C shift correlation (H–C COSY), the preferred experimental conditions are as follows: first, we begin with distinguishing between the pulse excitations of *A* and *X* spins (Figure 5-5a). Because of the large difference in

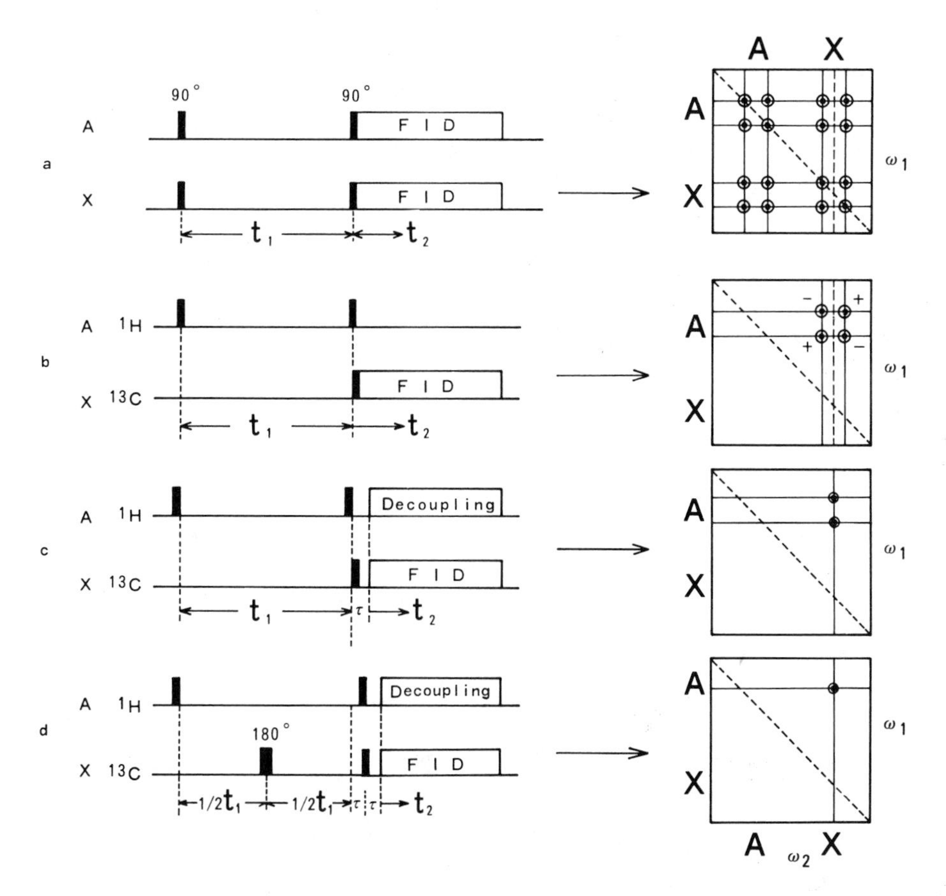

Figure 5-5. Transformation of experimental schemes from homonuclear 2D to heteronuclear 2D. An example of the transformation from COSY to H–C COSY (^1H–^{13}C shift correlation). (a) The COSY for an *AX* spin system. (b) Selective excitation of the *A* spin and selective detection of the *X* spin. (c) ω_2 decoupling in the detection period (t_2). (d) ω_1 decoupling in the evolution period (t_1) added to the ω_2 decoupling.

resonance frequencies of heterospins, this is accomplished quite easily. The normal COSY experiment in a heteronuclear system is shown in Figure 5-5a. Next, the first pulse excitation is restricted to nucleus *A* and the detection to nucleus *X*. The four split 2D peaks of a *AX* crosspeak appear as shown in Figure 5-5b. Here, it is important to recognize the difference in the sign of split peaks composing the crosspeak. The neighboring four peaks have mutually opposite signs. To avoid this splitting by *J* coupling in ^{13}C NMR, ^1H decoupling usually is employed. For the following reason, ^1H decoupling should start after some delay of the second pulse (Figure 5-5c); otherwise the neighboring peaks will overlap with opposite signs and so will cancel. The delay time is determined by the amount of *J* coupling as $\tau = (1/4J) \sim (1/2J)$. To decouple *J* coupling over the t_1 duration, a selective 180° pulse should be applied only to the *X* spin, leading to the focusing of *J*-modulated precession of the $A_{\pm}$spin components. Here some waiting time should also be inserted before the second pulse is applied, because at the refocusing point the $A_{\pm}$ spin components direct in parallel with the same phase. As indicated in Figure 5-3, $A_{\pm}$ spin components

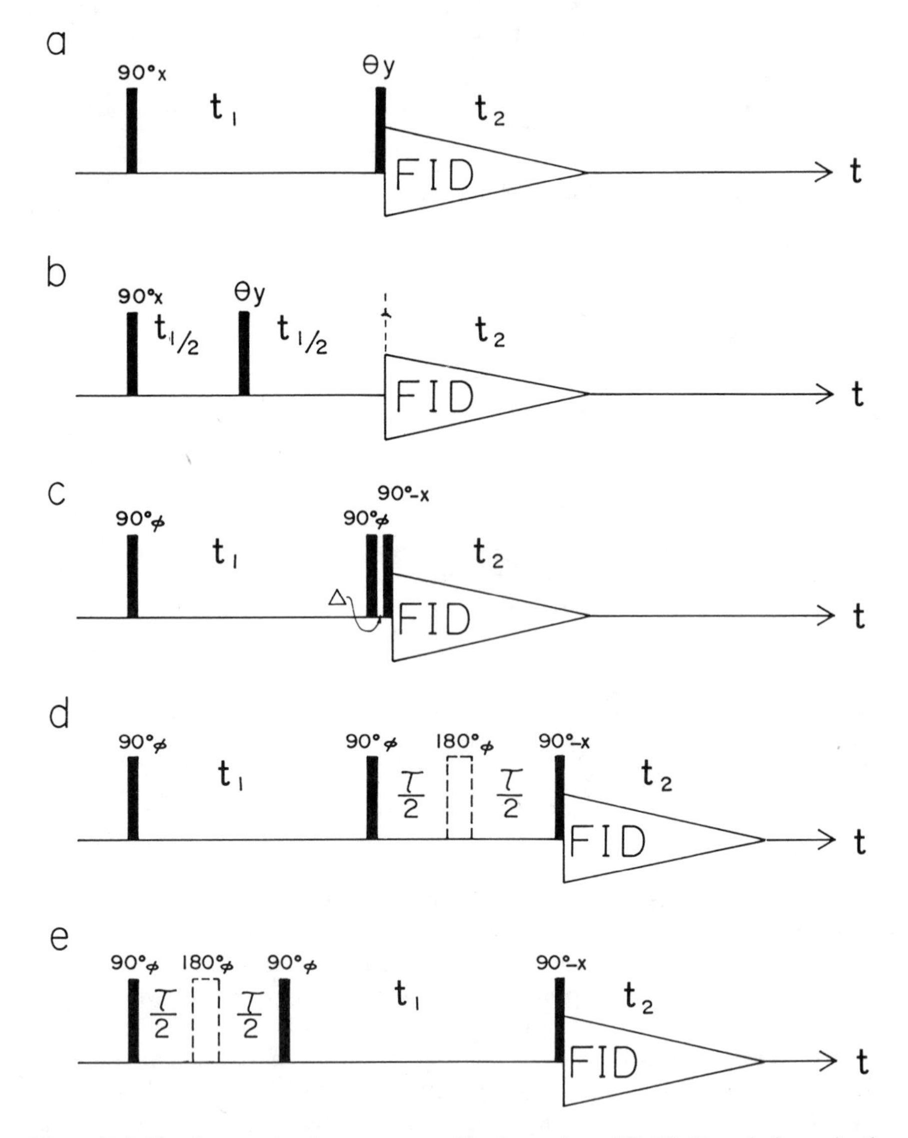

Figure 5-6. Fundamental pulse sequences for the various 2D NMR techniques in the homonuclear case. (a) COSY ($\theta_y = 90°$), (b) SECSY ($\theta_y = 90°$), 2DJ ($\theta_y = 180°$), (c) MQT-filter COSY, (d) NOESY, relay COSY, (e) MQT coherence. The subscript ø means phase cycling. The 180° pulse in schemes (d) and (e) is mandatory to perform 2D phase-sensitive detection for obtaining a pure phase.

should be in perfect antiphase in order to maximize the magnetization transfer from A to X by the 90° pulse. With these two decoupling procedures, H–C COSY is completed as shown in Figure 5-5d.

A few other ideas on deriving variations of 2D NMR from COSY have been proposed. Here only their experimental schemes are considered to show the relationship between the basic and derivative techniques. In Figure 5-6 typical pulse sequences are illustrated for the seven important classes of 2D NMR: COSY,[1] SECSY,[6] 2DJ,[3-5] MQT (multiple quantum transition) filter,[8] relay COSY,[9,10] NOESY[11] (nuclear Overhauser enhancement spectroscopy), and MQT coherence.[12] If these pulse sequences are combined with various selective techniques of excitation, detection, and decoupling, they are easily extended to heteronuclear 2D NMR as shown in Figure 5-5. The same pulse sequence is utilized in different 2D techniques. For example, sequence b corresponds either to SECSY or to 2DJ, depending on the flip angle of the second pulse. As explained in Figure 5-4, the essential difference between SECSY and 2DJ spectra lies in the intensities of 2D peaks. The 2DJ technique is simpler because it gives the same number of 2D peaks as does the 1D spectrum, but it contains a smaller volume of information as compared with SECSY.

Another example is sequence d, where NOESY and relay COSY are represented. Depending on whether interest is in the z component or in the x, y component after the second pulse, one of the two can be selected by the third read pulse by employing a pertinent combination of rf phases. If interest lies in the cross-relaxation of magnetizations of the z component during τ, NOESY results. If interest lies in the one additional transfer of magnetizations evolved during τ, relay COSY results by the third pulse. In sequences d and e, insertion of 180° pulse at the midpoint of the duration τ helps to present the 2D spectra in the phase-sensitive mode. The pulse sequence c of MQT filter is related to both COSY and MQT coherence because intensities of 2D peaks in COSY spectra are modified depending on the filtered MQT coherence. DQT (double quantum transition) filter, which can be obtained with the four rf phases 0°, 90°, 180°, and 270°, is very useful. Addition–subtraction (add–sub) of the resulting data eliminate singlets or singlet-like strong 2D peaks in the autopeaks and exaggerates the informative crosspeaks.

Information Filters and Partial Structures

Two-dimensional NMR techniques can visualize the topologic network of the interacting spin system through J couplings or cross-relaxations. The method of presenting topologic information varies from one technique to another, depending on the experimental schemes employed, as mentioned in the preceding section. This variety of techniques can be regarded as a kind of information filter to elucidate molecular structures. For example, if (with the aid of the combination of COSY and NOESY) a four-spin system can be identified

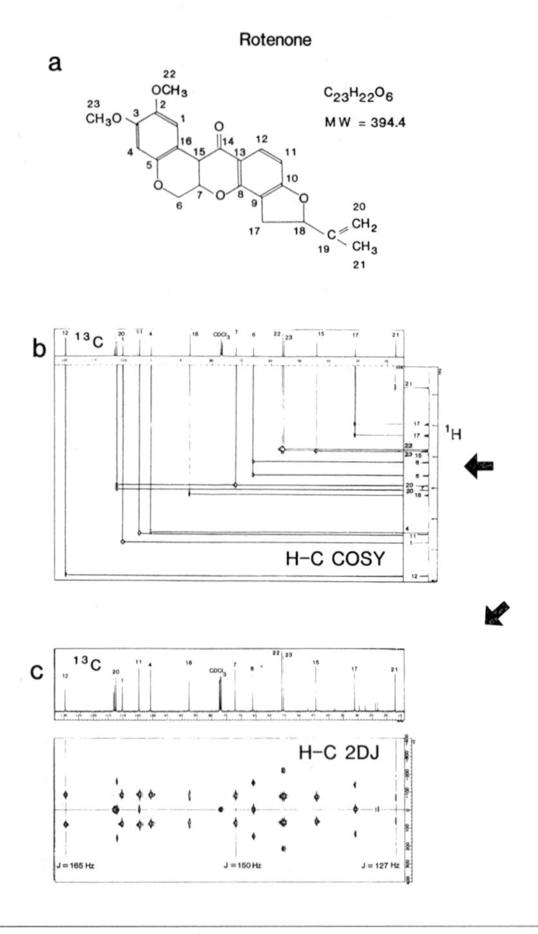

interacting with a three-spin system only through dipole–dipole interaction, we can safely assume that a partial structure consists of four nuclei and three nuclei mutually connected by some NMR-inactive nucleus (or nuclei). This simple correspondence between partial structures and spin systems renders 2D NMR a promising technique for structure elucidation. However, there are several exceptions to this rule that occur when coupling between nuclei belonging to the same spin system is very weak. Fortunately, NMR-active nuclei, such as ^{1}H, ^{13}C, ^{15}N, and ^{31}P, are the major constituents of organic molecules and biomolecules. What follows is a quick survey of what kinds of 2D information filters identify the partial structures. A medium-sized organic molecule, rotenone, is chosen as an illustrative example (Figure 5-7a). In Figure 5-7 various 2D spectra obtained for 0.2 M rotenone solutions are shown. They are arranged to illustrate the manner in which the basic 2D NMR (COSY) has evolved into other 2D techniques. The change of 2D spectral patterns can be traced, for example, with evolution to SECSY (Figure 5-7h) from COSY (Figure 5-7e) and with derivation to MQT filter (Figure 5-7d) from COSY.

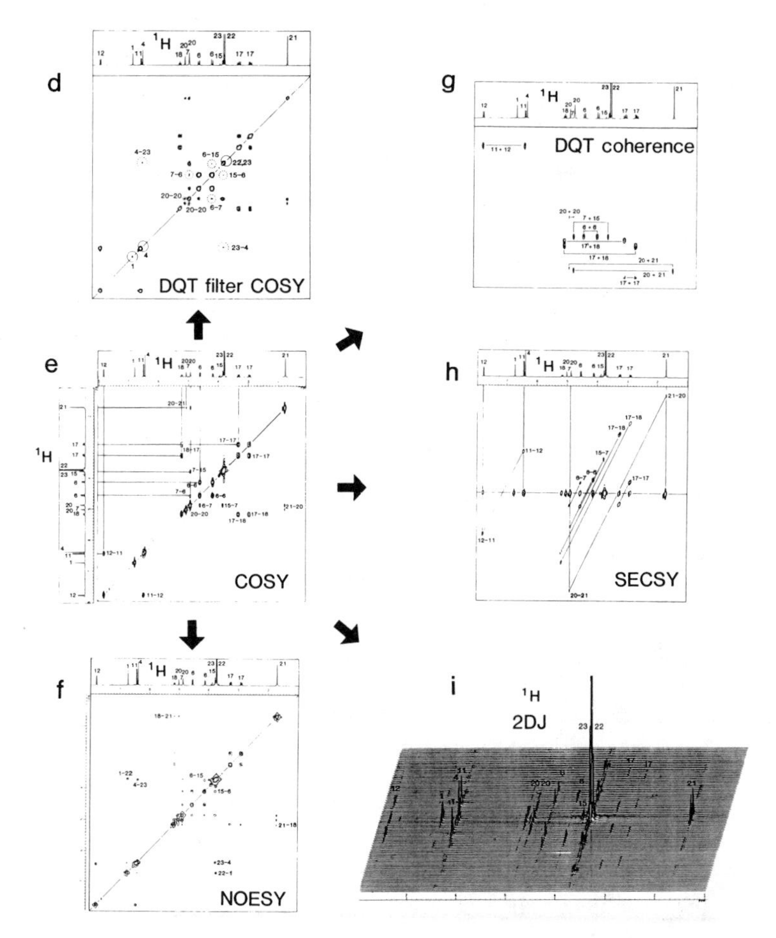

Figure 5-7. Two-dimensional NMR spectra for medium-size of organic molecule, rotenone. (a) Molecular structure of rotenone; (b)–(i) 2D NMR spectra obtained with various 2D techniques noted in the figure. All the spectra were taken with a 400-MHz NMR (GX-400, JEOL) attached with new pulse programmer PGX-300. The data matrix sizes utilized for the original time domain data were $4k \times 128$ to $4k \times 256$.

With COSY, SECSY, and DQT-filter COSY, various spin systems are identified for proton nuclei. The four resonances designated 6, 7, 7, and 15 constitute a well-coupled spin system. Protons 17, 17, and 18 constitute a three-spin system, and two protons, 11 and 12, constitute an isolated two-spin system. Two protons of 20 and methyl protons of 21 constitute a three-spin-like system because the three methyl protons behave as one nucleus. DQT filter COSY provided useful information to help identify the spin connections near the diagonal line (6–15, 20–20) and the long-range couplings (6–15, 4–23). When the same phase cycling and add-sub detection as that utilized in DQT-filter COSY is applied to MQT coherence, we obtain DQT coherence as shown in Figure 5-7g. In the ω_1 direction the frequency sum of two coupled protons develops at their chemical shift positions. The spin connections, therefore, appear horizontally. There is a simple transformation rule between MQT coherence and MQT filter COSY; thus, both contain identical information. In a crowded region that contains many resonances, however, the difference between their respective manners of presenting ω_1 frequencies may render the application of one technique preferable to another.

If we combine the above results with the information obtained with H–C COSY (Figure 5-7b) and H–C 2DJ (Figure 5-7c) regarding ^{13}C–^{1}H connections, partial structures including C and H are distinguished much better. The H–C COSY technique elucidates the direct connection of carbons and protons and H–C 2DJ informs us of the number of protons in the connection. One can easily follow the direct H–C connections and proton numbers with the two 2D spectra shown in Figures 5-7, b and c.

Protons that are spatially close but very weakly J coupled often do not get connected. For example, two methyls, 22 and 23, display only sharp singlets. In order to relate these protons to the partial structure of the same aromatic ring, NOESY must be employed (Figure 5-7f). The NOESY spectrum in Figure 5-7f successfully identifies two informative connections 1–22, 4–23. Another unique connection in the 2D spectrum is 18–21, which is difficult to identify by only COSY or MQT filter. In NOESY we often see the contamination of COSY response. For example, the 1-4 crosspeak observed in NOESY may arise from the survived J crosspeak with long-range coupling. Except for quaternary carbons and oxygens, the combination of eight different 2D spectra shown in Figure 5-7 elucidates the following partial structures: 22–1–4–23, 15–7–6, 11–12, and 17–18–20–21. In rotenone almost half of the carbon atoms are quaternary. Therefore, to finalize the overall structure, we must assign those quaternary carbons to the various partial structures and define the linkages that connect the partial structures involved.

Molecular Skeleton with Quaternary Carbons

The skeletons of almost all organic molecules are composed of carbon atoms (carbon skeleton). The interatomic connections between partial structures are often made through quaternary carbons. In our example, rotenone, except for the six oxygens, all of the backbone connections are made by carbons. Among them ten carbons exist as quaternary carbons that connect aromatic rings, olefins, and oxymethylenes. As mentioned in the preceding section, ^{1}H–^{1}H long-range couplings transmitted through quaternary carbons often enables the backbone connection to be followed (4–23). However, such good fortune cannot always be expected. Accordingly, more general methods using special classes of 2D techniques are considered next; these include ^{13}C–^{13}C DQT coherence, incredible natural abundance double quantum transfer experiment

(INADEQUATE),[13] pseudofiltered remote coupling H–C COSY,[14] and selective H–C 2DJ.[15,16]

Two-dimensional INADEQUATE is a special application of selective detection of DQT coherence of ^{13}C spin systems. The natural abundance of ^{13}C is about 1.1%. This leads to the conclusion that about 0.01% of adjacent carbon pairs are simultaneously ^{13}C and 0.0001% of adjacent carbon triplets are simultaneously ^{13}C on the carbon skeleton. Therefore, if SQT (single quantum transition) originating from isolated ^{13}C nuclei can be eliminated and only DQT originating from ^{13}C pairs selected, then 2D spectra that connect nearest neighbors of the carbon skeleton can be obtained. In the 2D spectra, DQT from spin systems larger than two nuclei can be neglected because of their low probability of occurrence, as mentioned. This permits unambiguous, step-by-step tracing of carbon skeleton. A 2D INADEQUATE spectrum for rotenone is shown in Figure 5-8. Ten quaternary carbons that were neglected in the preceding section are now equally weighted for the purpose of structural elucidation. Starting from known methyl or methylene carbons, the carbon skeleton can be traced easily. For example, starting from a methyl carbon, 21, the 21–19 connection is readily identified. Then, one of the two branches, 19–18 and 19–20 can be identified easily because carbons 18 and 20 are already assigned. Another skeleton, 18–17–9–10–11–12, is also identified easily because the 12–11 pair of carbons can be recognized. Carbons 9 and 13, both of which are connected to carbon 8, show similar chemical shifts and, hence, both connections overlap. Starting from carbon 13, then the analysis proceeds to carbon 14 and to two branches, 14–15–7–6 and 14–15–16. The final aromatic ring connections are also traced easily as 16–1–2 and 16–5–4. The 2–3 connection is obscured. To confirm the carbon skeleton, one can also start from carbon 6 without fear of contradiction. Finally, together with the total assignments of individual protons and carbons, the molecular structure itself can be elucidated, except for connections resulting from oxygen bonding. Two oxygens of the oxymethyls are recognized because of the lower chemical shifts of the $—OCH_3$ methyl protons. The carbonyl carbon, 14, is also recognized by its downfield chemical shift. The remaining three oxygens are known because of the tendency of lowering the chemical shifts of nearest neighboring carbons. There are, however, no *a priori* preferable carbon connections through oxygen with the oxygen-bonded carbons 5, 6, 7, 8, 10, and 18. Therefore, the structure suggested in Figure 5-7a cannot be determined uniquely by all of the knowledge obtained up to now. The connection between 5 and 18 through oxygen might be feasible, for example. It is necessary to invoke quantum chemical considerations if it is to be concluded that the carbon connection through oxygen, such as 5 to 18, 6 to 10, or 6 to 8, is not normal.

The H–C COSY technique is often useful for assigning the carbon skeleton through the use of quaternary carbons when remote coupling between ^{1}H and ^{13}C is utilized. Quaternary carbons 2 and 3 may be identified by their remote coupling to protons 1 and 4. The pseudofilter technique used to determine the small H–C J couplings is combined with H–C COSY, thereby eliminating

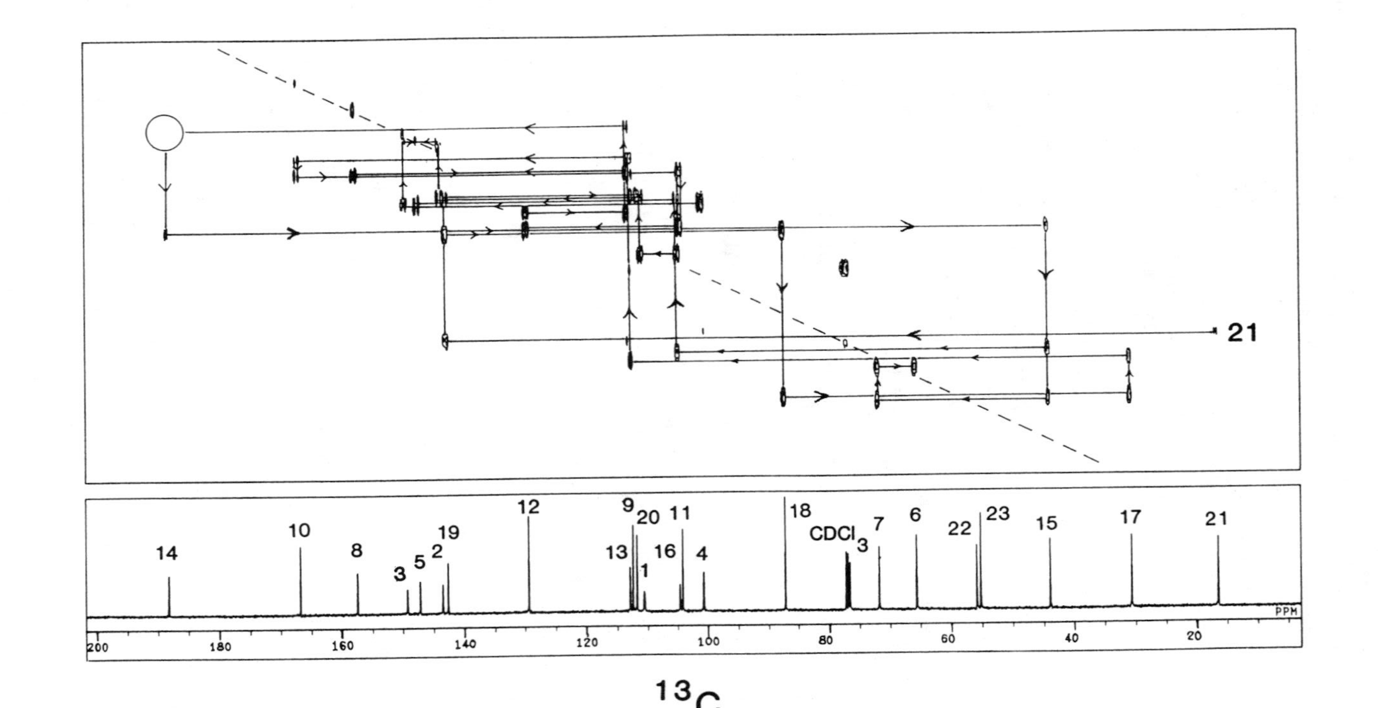

Figure 5-8. The 2D INADEQUATE (^{13}C–^{13}C DQT coherence) for rotenone. One partner of the 13–14 connection is missing as shown by a circle. The 400-MHz NMR spectrum was recorded for a 0.5 M $CDCl_3$ solution of the sample. The original data matrix was $4k \times 256$.

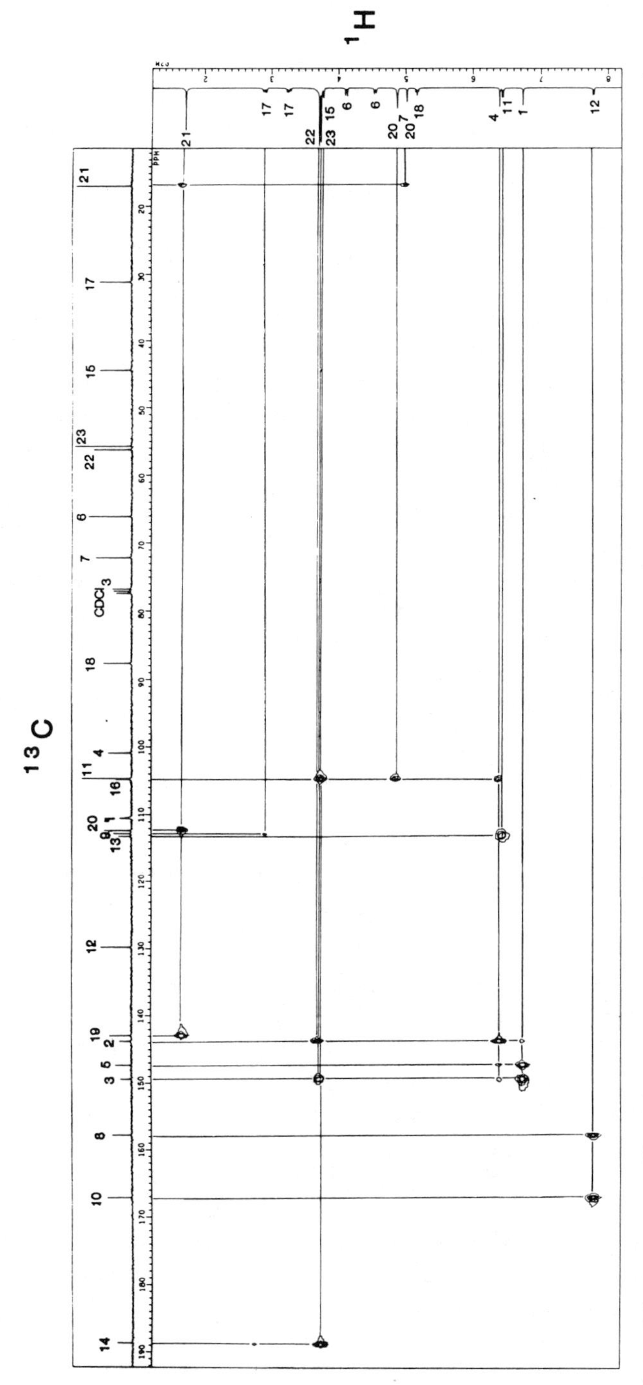

Figure 5-9. Remote coupling H–C COSY with pseudofilter (quaternary carbon selection) for rotenone. A 0.2 M $CDCl_3$ solution of the sample was used. The original data matrix was $4k \times 256$.

carbon resonances that are bonded directly to proton(s).[14] An example of the 2D technique for rotenone is shown in Figure 5-9. By using *a priori* ^{13}C NMR knowledge, ie, that the carbonyl group resonates at very lowfield and oxygen bonded carbons show lower chemical shifts than do carbons that are surrounded only by protons and carbons, the assignments shown in Figure 5-9 could be completed on the basis of the analysis shown in the preceding section. Of course, long-range H–C coupling information is not enough to permit all of the individual assignments to be made. For example, assignments of protons 1 and 4 and carbons 1, 2, 3, and 4 are still ambiguous because of pseudosymmetry between 3–4–5 and 2–1–16 nuclei and lack of long-range couplings to the remaining fragments. The long-range H–C COSY of Figure 5-9 was used specifically to determine individual assignments of 11 and 12 protons. The two oxygen-bonded carbons, 8 and 10, which are coupled remotely to proton 12, should show lower chemical shifts than should the other two carbons 9 and 13, which are coupled remotely to proton 11. The two protons, 11 and 12, in Figure 5-9 are thereby assigned.

By applying selective decoupling[16] or selective 180° pulse[15] for protons in H–C 2DJ, H–C remote coupling can be observed between the selectively irradiated proton and the coupled carbons. The rf power of the proton irradiation should be much smaller compared with the power to decouple large direct H–C coupling (about 150 Hz); otherwise the remote coupling cannot be accomplished selectively. The technique has been utilized practically in structural eludication of a natural organic molecule, oxyrapentyn.[16] From the point of view of sensitivity, the latter two techniques are better than 2D INADEQUATE, although the carbon skeleton is determined less ambiguously with the C–C DQT coherence.

Future Developments

There are two difficult problems that should be addressed in an effort to improve the 2D techniques described. One is the existence of short spin–spin relaxation times, T_2 which makes the NMR signal relax before the data acquisition of 2D development. Another problem is to determine the structure of molecular skeletons that are composed of heteronuclei other than carbon, eg, oxygen, nitrogen, or phosphorus. The T_2 problem is especially serious in 2D application of organic molecules with molecular weight larger than 1000 where the relaxation rate $1/T_2$, often becomes larger than $2\pi J$ for the remote H–C coupling. One remedy to avoid rapid relaxation of transverse magnetizations is to maximize the resultant 2D signal by finding the smallest possible delay, τ, or deliberately abandoning the selective decoupling utilized in H–C COSY to make $\tau=0$. Heteronuclear NOESY (for example, H–C NOESY) might provide an alternative to the remote H–C COSY if the cross-relaxation rate between ^{13}C and ^{1}H is larger than their remote coupling constant.

With regard to oxygen and nitrogen nuclei, fortunately NMR active isotopes are available, eg, ^{17}O ($I = 5/2$) and ^{15}N ($I = 1/2$). Selective labeling of ^{17}O has been utilized already. The use of natural abundance ^{15}N NMR is also now under consideration for heteronuclear 2D NMR. Phosphorus-31 is abundant and is a rather highly sensitive nucleus, and so there is no difficulty in applying ^{31}P heteronuclear 2D NMR techniques. The unification of the 2D techniques raised in the last two sections, together with future developments of 2D techniques here mentioned, may make possible the unambiguous determination of the structures of a number of organic molecules. Before the final goal, automatic analysis of chemical structures by 2D unification, is reached, however, many details concerning practical applications and standardization of measurement conditions should be established.

Appendix: Brief Survey of Applications of Various 2D Techniques in Biologic Systems

In this Appendix recent studies of applications of 2D techniques to biologically important substances are reviewed briefly. Our attention is focused on the usefulness of 2D techniques, singly and in combinations.

Proteins

Extensive studies of individual resonance assignments with COSY, SECSY, NOESY, and 2DJ for small proteins (mainly basic pancreatic trypsin inhibitor, BPTI) have been performed by the Wüthrich group.[17-32] Especially the backbone protons NH and $C_{\alpha}H$ have been systematically assigned making use of the knowledge of the primary sequence of the protein concerned.[33] The application of various 2D techniques to protein NMR has recently been reviewed comprehensively.[34] For BPTI, NMR parameters for the modified and unmodified protein and the calculation of comformation changes thus introduced were compared.[35,36] More sophisticated 2D NMR techniques, such as DQT filter or relay COSY, have been applied to proteins.[37-39] The NOESY technique has been utilized not only to elucidate cross-relaxation phenomena but also to observe chemical exchanges that occur in proteins, eg, enzyme kinetics.[40] Very few heteronuclear 2D NMR studies have been done on proteins,[41,42] although this technique is expected to find wide application to protein studies because of its great ability to resolve many overlapping peaks in complex NMR spectra. In the field of protein–DNA interaction, several repressor protein studies have been performed using 2D NMR techniques.[32,43-45]

Peptides (Hormones, Antibiotics and Others)

For biologically active oligopeptides and peptide derivatives, various 2D homonuclear and heteronuclear 2D NMR techniques have been applied. Systematic applications of combined 2D techniques are in progress in the Kessler group (see Chapter 6) for the structural elucidation of cyclic peptides.[46-50] In order to determine the backbone structure, remote coupling H–C COSY has been utilized actively.[50] Initial applications of simple 2D NMR techniques to peptide chemistry that have involved 2DJ, COSY, and SECSY, have been concerned primarily with resonance assignments.[51-53] The use of 2D NMR is now extended to include NOESY[54] and DQT coherence.[55]

Oligonucleotides

An increasing number of studies involving structural transitions of oligonucleotides and protein–DNA interactions have been reported. Important applications of 2D techniques have also involved individual assignments of proton and carbon resonances.[56-58] A quite similar strategy for the systematic sequential resonance assignments of nucleotide residues at sugar and base positions, using NOESY and COSY, has been proposed by Reid's group,[59,60] Kaptein's group,[61,62] and the other groups.[63,64] The NOESY technique also has been utilized to study DNA conformation, including B–Z transition, by Kearns' group[65-67] and by others.[68] The interaction between DNA and antibiotics, eg, actinomycin D, has been studied by 2D NMR.[69,70] The NMR-active nucleus ^{31}P, unique in DNA, should be utilized extensively in the future in heteronuclear 2D NMR techniques.[71,72]

tRNAs

The application of 2D to tRNA has been scant as compared with oligonucleotides.[73,74] This is because of the lack of a uniform secondary structure, which is fully appreciated when making sequential assignments of DNA. Conformation studies in solution, however, would be the next good target for study using combined 2D techniques, such as NOESY, heteronuclear NOESY,[75] and heteronuclear MQT.[76]

Saccharides

Overlapping of many spin systems is well resolved by using COSY, NOESY, or MQT-filter COSY. Application of 2D to saccharides and polysaccharides is appropriate[77] because it can resolve complicated overlapping signals in the proton NMR resonances of sugar rings. Use of COSY, SECSY, and 2DJ for resonance assignments has been made for oligosaccharides.[78-81] Resonance assignments and structural determination with COSY and NOESY were carried out for a tetrasaccharide[82] and for an oligosaccharide from glycoprotein.[83] More sophisticated 2D techniques with ^{1}H–^{19}F COSY[84] and ^{1}H–^{1}H–^{13}C relay COSY[85] also have been applied.

Other Natural Products

Two-dimensional INADEQUATE was specifically utilized to distinguish one of the two possible structures of a steroid photodimer.[86] Combined use of various 2D techniques was motivated to assign proton resonances of such natural products as α-biotin[87] and retinal.[88] Combinations of 2D INADEQUATE, H–C COSY, and COSY have been utilized to elucidate the structures of several steroids[89] and of a mutagen.[90] ^{13}C–^{13}C COSY and H–C COSY techniques are again important tools for making structural assignments,[91,92] and NOESY has been utilized for the structural elucidation of a coenzyme (methanopterin),[93] a toxin (taleromycin B),[94] and a macrocyclic (cytochalasin B).[95]

Finally reference should be made to a very intriguing paper that deals with the use of 2D NMR in combination with an interactive computer program for solving structural problems.[96] This type of work suggests that 2D techniques will find increasing applications to solving complex structural problems in future years.

References

1. Aue,, W. P.; Bartholdi, E.; Ernst, R. R. *J. Chem. Phys.* **1976**, *64*, 2229.
2. Sørensen, O. W.; Eich, G. W.; Levitt, M. H.; Bodenhausen, G.; Ernst, R. R. *Prog. NMR Spectrosc.* **1983**, *16*, 163.
3. Aue, W. P.; Karhan, J.; Ernst, R. R. *J. Chem. Phys.* **1976**, *64*, 4226.
4. Bodenhausen, G.; Freeman, R.; Niedermeyer, R.; Turner, D. L. *J. Magn. Resonance* **1976**, *24*, 291.
5. Nagayama, K.; Bachmann, P.; Wüthrich, K.; Ernst, R. R. *J. Magn. Resonance* **1978**, *31*, 133.
6. Nagayama, K.; Anil Kumar; Wüthrich, K.; Ernst, R. R. *J. Magn. Resonance* **1980**, *40*, 321.
7. Sørensen, O. W.; Rance, M.; Ernst, R. R. *J. Magn. Resonance* **1984**, *56*, 527.
8. Piantini, U.; Sørensen, O. W.; Ernst, R. R. *J. Am. Chem. Soc.* **1982**, *104*, 6800.
9. Eich, G.; Bodenhausen, G.; Ernst, R. R. *J. Am. Chem. Soc.* **1982**, *104*, 3731.
10. Bolton, P. H.; Bodenhausen, G. *Chem. Phys. Lett.* **1982**, *89*, 139.
11. Jeener, J.; Meier, B. H.; Bachmann, P.; Ernst, R. R. *J. Chem. Phys.* **1979**, *71*, 4546.
12. Wokaun, A.; Ernst, R. R. *Chem. Phys. Lett.* **1977**, *52*, 407.
13. Bax, A.; Freeman, R.; Kempsell, S. P. *J. Am. Chem. Soc.* **1980**, *102*, 4849.
14. Kurihara, N.; Kamo, O.; Umeda, M.; Sato, K.; Hyakuna, K.; Nagayama, K. *J. Magn. Resonance* **1985**. In press.
15. Bax, A.; Freeman, R. *J. Am. Chem. Soc.* **1982**, *104*, 1099.
16. Seto, H.; Furihata, K.; Otake, N.; Itoh, Y.; Takahashi, S.; Haneishi, T.; Ohuchi, M. *Tetrahedron Lett.* **1984**, *25*, 337.
17. Nagayama, K.; Wüthrich, K. *Eur. J. Biochem.* **1981**, *114*, 365.
18. Wagner, G.; Anil Kumar; Wüthrich, K. *Eur. J. Biochem.* **1981**, *114*, 375.
19. Nagayama, K.; Wüthrich, K. *Eur. J. Biochem.* **1981**, *115*, 653.
20. Brown, L. R.; Wüthrich, K. *Biochim. Biophys. Acta* **1981**, *647*, 95.
21. Anil Kumar; Wagner, G.; Ernst, R. R.; Wüthrich, K. *J. Am. Chem. Soc.* **1981**, *103*, 3654.
22. Brown, L. R.; Braun, W.; Anil Kumar; Wüthrich, K. *Biophys. J.* **1982**, *37*, 319.

23. Wüthrich, K.; Wider, G.; Wagner, G.; Braun, W. *J. Mol. Biol.* **1982**, *155*, 311.
24. Billeter, M.; Braun, W.; Wüthrich, K. *J. Mol. Biol.* **1982**, *155*, 321.
25. Wagner, G.; Wüthrich, K. *J. Mol. Biol.* **1982**, *155*, 347.
26. Wider, G.; Lee, H.-K.; Wüthrich, K. *J. Mol. Biol.* **1982**, *155*, 367.
27. Arseniev, A. S.; Wider, G.; Joubert, F. J.; Wüthrich, K. *J. Mol. Biol.* **1982**, *159*, 323.
28. Keller, R. M.; Baumann, R.; Hunzinger-Kwik, E.-H.; Joubert, F. J.; Wüthrich, K. *J. Mol. Biol.* **1983**, *163*, 623.
29. Štrop, P.; Wider, G.; Wüthrich, K. *J. Mol. Biol.* **1983**, *166*, 641.
30. Štrop, P.; Čechova, D.; Wüthrich, K. *J. Mol. Biol.* **1983**, *166*, 669.
31. Marion, D.; Wüthrich, K. *Biochem. Biophys. Res. Commun.* **1983**, *113*, 967.
32. Zuiderweg, E. R. P.; Kaptein, R.; Wüthrich, K. *Proc. Natl. Acad. Sci. USA* **1983**, *80*, 5837.
33. Wüthrich, K.; Wider, G.; Wagner, G.; Braun, W. *J. Mol. Biol.* **1982**, *155*, 311.
34. Wider, G.; Macura, S.; Anil Kumar; Ernst, R. R.; Wüthrich, K. *J. Magn. Resonance* **1984**, *56*, 207.
35. Nagayama, K.; *Adv. Biophys.* (Tokyo) **1981**, *14*, 139.
36. Yoshioki, S.; Abe, H.; Noguti, T.; Gō, N.; Nagayama, K. *J. Mol. Biol.* **1983**, *170*, 1031.
37. Wagner, G.; Zuiderweg, R. P. *Biochem. Biophys. Commun.* **1983**, *113*, 854.
38. Wagner, G. *J. Magn. Reson.* **1983**, *55*, 151.
39. Boyd, J.; Dobson, C. M.; Redfield, C. *J. Magn. Resonance* **1983**, *55*, 170.
40. Balaban, R. S.; Ferretti, J. A. *Proc. Natl. Acad. Sci. USA* **1983**, *80*, 1241.
41. Chan, T. M.; Markley, J. L. *Biochemistry* **1983**, *22*, 5996.
42. Kojiro, C. L.; Markley, J. L. *FEBS Lett.* **1983**, *162*, 52.
43. Ribeiro, A. A.; Wemmer, D.; Bray, R. P.; Wade-Jardetzky, N. G.; Jardetzky, O. *Biochemistry* **1981**, *20*, 818.
44. Ardnt, K. T.; Boschelli, F.; Cook, J.; Takeda, Y.; Tecza, E.; Lu, P. *J. Biol. Chem.* **1983**, *258*, 4177.
45. Weiss, M. A.; Karplus, M.; Patel, D. J.; Sauer, R. T. *J. Biomol. Struct. Dyn.* **1983**, *1*, 151.
46. Kessler, H.; Hehlein, W.; Schuck, R. *J. Am. Chem. Soc.* **1982**, *104*, 4534.
47. Kessler, H.; Bermel, W.; Friedrich, A.; Krack, G.; Hull, W. E. *J. Am. Chem. Soc.* **1982**, *104*, 6297.
48. Kessler, H.; Bernd, M.; Kogler, H.; Zarbock, J.; Sørensen, O. W.; Bodenhausen, G.; Ernst, R. R. *J. Am. Chem. Soc.* **1983**, *105*, 6944.
49. Kessler, H.; Oschkinat, H.; Sørensen, O. W.; Kogler, H.; Ernst, R R. *J. Magn. Resonance* **1983**, *55*, 329.
50. Kessler, H.; Griesinger, C.; Zarbock, J.; Loosli, H. R. *J. Magn. Reson.* **1984**, *57*, 331.
51. Kobayashi, Y.; Kyogoku, Y.; Emura, J.; Sakakibara, S. *Biopolymers* **1981**, *20*, 2021.
52. Clayden, N. J.; Inagaki, F.; Williams, R. J. P.; Morris, G. A.; Tori, K.; Tokura, K.; Miyazawa, T. *Eur. J. Biochem.* **1982**, *123*, 127.
53. Krishna, N. R.; Heavner, G. A.; Vaughn, J. B. Jr. *J. Am. Chem. Soc.* **1983**, *105*, 6930.
54. Cutnell, C. D. *J. Am. Chem. Soc.* **1982**, *104*, 362.
55. Macura, S.; Kumar, N. G.; Brown, L. R. *Biochim. Biophys. Res. Commun.* **1983**, *117*, 486.
56. Kan, L-S.; Cheng, D. M.; Cadet, J. *J. Magn. Reson.* **1982**, *48*, 86.
57. Frechet, D.; Cheng, D. M.; Kan, L-S.; Ts'o, P. O. P. *Biochemistry* **1983**, *22*, 5194.
58. Pardi, A.; Waslker, R.; Rapoport, H.; Wider, G.; Wüthrich, K. *J. Am. Chem. Soc.* **1983**, *105*, 1652.
59. Hare, D.; Wemmer, D. E.; Chou, S-H.; Drobny, G.; Reid, B. R. *J. Mol. Biol.* **1983**, *171*, 319.
60. Wemmer, D. E.; Chou, S-H.; Hare, D. R.; Reid, B. R. *Biochemistry* **1984**, *23*, 2262.
61. Scheek, R. M.; Russo, N.; Boelens, R.; Kaptein, R.; van Boom, J. J. *J. Am. Chem. Soc.* **1983**, *105*, 2914.
62. Scheek, R. M.; Bolens, R.; Russo, N.; van Boom, J. H.; Kaptein, R. *Biochemistry* **1984**, *23*, 1371.
63. Weiss, M. A.; Patel, D. J.; Sauer, R. T.; Karplus, M. *Proc. Natl. Acad. Sci. USA* **1984**, *81*, 130.
64. Broido, M. S.; Zon, G.; James, T. L. *Biochem. Biophys. Res. Commun.* **1984**, *119*, 663.
65. Broido, M. S.; Kearns, D. R. *J. Am. Chem. Soc.* **1982**, *104*, 5207.
66. Feigon, J.; Wright, J. M.; Leupin, W.; Denny, W. A.; Kearns, D. R. *J. Am. Chem. Soc.* **1982**, *104*, 5540.
67. Assa-Munt, N.; Kearns, D. R. *Biochemistry* **1984**, *23*, 791.
68. Feigon, J.; Wang, A. H-J.; van der Marel, G. A.; van Boom, J. H.; Rich, A. *Nucl. Acids. Res.* **1984**, *12*, 1243.

69. Brown, S. C.; Mullis, K.; Levenson, C.; Shater, R. H. *Biochemistry* **1984**, *23*, 403.
70. Reid, D. G.; Doddrell, D. M.; Fox, K. R.; Salisbury, S. A.; Williams, D. H. *J. Am. Chem. Soc.* **1983**, *105*, 5945.
71. van Divender, J. M.; Hutton, W. C. *J. Magn. Reson.* **1982**, *48*, 272.
72. Jacobson, L. *J. Magn. Resonance* **1982**, *49*, 522.
73. Haasnoot, C. A. G.; Heerschap, A.; Hilbers, C. W. *J. Am. Chem. Soc.* **1983**, *105*, 5483.
74. Griffery, R. H.; Poulter, C. D.; Bax, A.; Hawkins, B. L.; Yamaizumi, Z.; Nishimura, S. *Proc. Natl. Acad. Sci. USA* **1983**, *80*, 5895.
75. Yu, C.; Levy, G. C. *J. Am. Chem. Soc.* **1983**, *105*, 6994.
76. Bax, A.; Griffery, R. H.; Hawkins, B. L. *J. Am. Chem. Soc.* **1983**, *105*, 7188.
77. Hall, L. D.; Morris, G. A.; Sukumar, S. *J. Am. Chem. Soc.* **1980**, *102*, 1745.
78. Bernstein, M. A.; Hall, L. D. *J. Am. Chem. Soc.* **1982**, *104*, 5553.
79. Bruch, R. C.; Bruch, M. D. *J. Biol. Chem.* **1982**, *257*, 3409.
80. Homars, S. W.; Dwek, R. A.; Fernandes, D. L.; Rademacher, T. W. *Biochim. Biophys. Acta* **1983**, *760*, 256.
81. Koerner, T. M. Jr.; Prestegard, J. H.; Demou, P. C.; Yu, R. K. *Biochemistry* **1983**, *22*, 2676.
82. Bothner-By, A. A.; Stephens, R. L.; Lee, J. *J. Am. Chem. Soc.* **1984**, *106*, 811.
83. Bhattacharyya, S. N.; Lynn, W. S.; Dabrowski, J.; Tauner, K.; Hull, W. E. *Arch. Biochem. Biophys.* **1984**, *231*, 72.
84. Card, P. J.; Reddy, G. S. *J. Org. Chem.* **1983**, *48*, 4734.
85. Bigler, P.; Anmann, W.; Richarz, R. *Org. Magn. Resonance* **1984**, *22*, 109.
86. Freeman, R.; Frenkiel, T.; Rubin, M. B. *J. Am. Chem. Soc.* **1982**, *104*, 5545.
87. Ikura, M.; Hikichi, K. *Org. Magn. Resonance* **1982**, *20*, 266.
88. Wernly, J.; Lauterwein, J. *Helv. Chim. Acta* **1983** *66*, 1576.
89. Bhacca, N. S.; Balandrin, M. F.; Kinghorn, A. D.; Frenkiel, T. A.; Freeman, R.; Morris, G. A. *J. Am. Chem. Soc.* **1983**, *105*, 2538.
90. Musmar, M. J.; Willcott, M. R. III; Martin, G. E.; Gampe, R. T. Jr.; Iwano, M.; Lee, M. L.; Hurd, R. E.; Johnson, L. F.; Castle, R. N. *J. Heterocycl. Chem.* **1983**, *20*, 1661.
91. Gampe, R. T. Jr.; Maktoob, A.; Weinheimer, A. J.; Martin, G. E.; Matson, J. A.; Willcott, M. R. III; Inners, R. R.; Hurd, R. E. *J. Am. Chem. Soc.* **1984**, *106*, 1823.
92. Ubukata, M.; Uzawa, J.; Isono, K. *J. Am. Chem. Soc.* **1984**, *106*, 2213.
93. van Beelen, P.; Stassen, A. P. M.; Bosh, J. W. G.; Vogels, G. D.; Guijt, W.; Haasnoot, C. A. G. *Eur. J. Biochem.* **1984**, *138*, 563.
94. Hutton, W. C.; Phillips, N. J.; Garden, D. W.; Lynn, D. G. *J. Chem. Soc., Chem. Commun.* **1983**, 864.
95. Garden, D. W.; Lynn, D. G. *J. Am. Chem. Soc.* **1984**, *106*, 1119.
96. Lindley, M. R.; Shoolery, J. N.; Smith, D. H.; Djerassi, C. *Org. Magn. Resonance* **1983**, *21*, 405.

6

CONFORMATIONAL ANALYSIS OF PEPTIDES BY TWO-DIMENSIONAL NMR SPECTROSCOPY*

Horst Kessler and Wolfgang Bermel

INSTITUTE OF ORGANIC CHEMISTRY
JOHANN WOLFGANG GOETHE-UNIVERSITÄT
D-6000 FRANKFURT 50
FEDERAL REPUBLIC OF GERMANY

Introduction

The significance of peptide conformations for their biological activity is now generally accepted. Knowledge of the conformational behavior is the molecular basis for understanding such phenomena as active transport, cell differentiation, and biological regulation via hormone–receptor interaction.[2–4]

Peptides and proteins consist of rather flexible chains or rings of amino acids. Ordering of structure is mainly introduced by weak intramolecular interactions (hydrogen bonding, hydrophobic interactions, etc.). In general these weak molecular forces depend upon the medium. For structure–activity discussions the receptor-side conformation is required, which in the first approximation is simulated by the conformation in solution. Although the most detailed structural picture is obtained by x-ray diffraction of single crystals, its application is restricted to crystalline samples and yields per se only the solid-state conformation. Especially in small peptides the conformation in solution may differ dramatically from those in the crystal. Some examples have been

*This is contribution number 34 in the series *Peptide Conformations*. For preceding paper, see Reference 1.

NMR in Stereochemical Analysis

published in which such conformational changes occurring when a crystal dissolves can be observed directly.[5,6] The difference in conformation of crystal and solution mainly results from stronger *intermolecular* interactions in the solid state than in solution. As an example, x-ray analysis of several cyclic hexapeptides has often shown only one or two intramolecular hydrogen bonds (β turns). These "transannular hydrogen bonds are not necessarily essential to conformational stability of the peptide ring"[7] in the crystal because the molecule is spanned in the rigid lattice forming strong *intermolecular* hydrogen bonds. Nuclear magnetic resonance spectroscopy, on the other hand, has provided evidence that up to three internal hydrogen bonds in cyclic hexapeptides can be formed in solution.[8]

The most suitable method of studying conformation in solution is NMR spectroscopy. Its outstanding relevance results from its ability to detect specific atoms (^{1}H, ^{13}C, ^{15}N) within a molecule and to indicate their surroundings via through-space and through-bond effects. As with any other spectroscopic method, the structural interpretation of the spectra depends upon the reliability of spectral assignments and the accuracy of the spectral parameters that contain the information. The introduction of two-dimensional NMR spectroscopy has revolutionized the extraction and assignment of spectral parameters. In this chapter some applications of these new methods to peptide spectra are demonstrated. Because of the large number of recent publications this discussion is restricted more or less to those that deal with peptides and proteins. New techniques are cited more as introduction for the reader, than as a chronological review.

Some Remarks on Peptide Conformation Analysis: The Conformational Homogeneity Problem

The high flexibility of peptides results mainly from rotational freedom about single bonds, such as N—C_α (bond angle ϕ) or C_α—CO (angle ψ), in the backbone as well as the side chains (for definition of rotational angles see Figure 6-1).

Rotation about ψ and ϕ requires less than 4 kcal mol^{-1}; therefore, rotamers have not been distinguished so far in the NMR spectrum. Only rotation about the amide bond (cis–trans isomerism) is slow on the NMR time scale.[10-12] With a typical barrier of 18–20 kcal mol^{-1} coalescence phenomena are observed in the range of 80–130°C in cases where both rotamers are almost equally populated. This barrier allows isolation of stable isomers not at room temperature, but at lower temperatures.[13] Conformational analysis of peptides must always consider the number of conformations in fast equilibrium. At present, there is no method of analyzing such a complex conformational mixture directly. Moreover, postulation of a so-called "mean conformation" is physically not meaningful.[2,14]

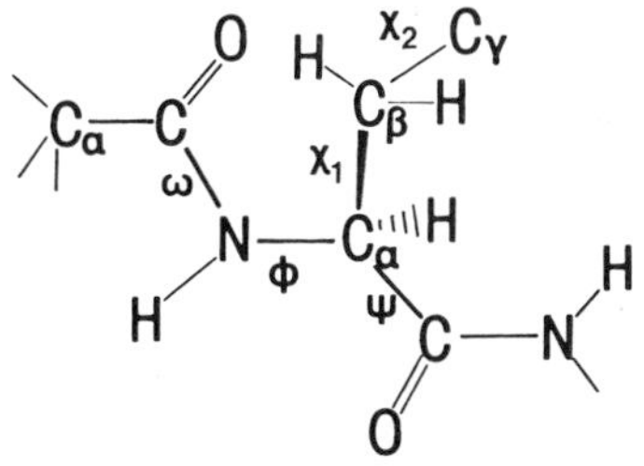

Figure 6-1. Definition of bond angles in peptides.[9] The values and signs of the angles formed by the backbone are determined by looking along the peptide chain from the N- to the C-terminal end. For definition of the χ angles see ref. 9.

Fortunately, there often exist several conformational restrictions. The different ground state energies of rotamers about both the CO—C_α and the N—C_α bonds lead to a preference of certain ψ and ϕ angles, which usually are represented by a Ramachandran plot (energy as function of ϕ and ψ).[15] In proteins a number of further secondary stabilizing interactions—hydrogen bonding, hydrophobic interactions, ionic forces—often lead to a strong restriction of flexibility. However recent NMR studies of small proteins [basic pancreatic trypsin inhibitor (BPTI), glucagon] have shown that the assumption of a rigid protein conformation is not always true.[16-18] Smaller peptides generally exhibit a higher flexibility. Backbone rigidity must be introduced via cyclization[3,4,19,20] to reduce the conformational manifold. Even in cyclic peptides, however, the first question in conformational analysis is to provide evidence of a dominance of one conformation in the equilibrium. Two conformations in slow exchange compared to the NMR time scale often also can be analyzed,[21] but generally more than two species lead to spectra that are too complicated even for modern techniques.

There exist several criteria for conformational homogeneity. None of them is stringent but, normally, in summing up the different arguments a relatively convincing picture of the flexibility can be obtained. Arguments of conformational homogeneity are:

1. Strong differentiation of spectral parameters of equal amino acids in the sequence (proton and/or carbon chemical shifts and coupling constants)
2. Strong differentiation in the NH chemical shift values and their temperature dependence of the different amino acid residues in the molecule
3. NH–C_αH vicinal coupling constants that differ greatly from the "mean value" of about 7.5 Hz
4. Strong splittings of the diastereotopic C_αH protons of glycine
5. Side chain fixation, which may be detected via strong differentiation of β protons in chemical shift and coupling constants (eg, in Phe, Tyr, or Trp residues)
6. Apparent independence of conformations from solvent and temperature (no

change of J values in different solvents; linearity of NH chemical shift variation with temperature)
7. Accordance of conformational conclusions obtained from NH temperature dependence and solvent titration (DMSO→$CDCl_3$) data.

Hence, careful conformational analysis requires the reliable check of as many of the above-mentioned parameters as can be obtained. This chapter considers only methods of obtaining spectra and their analysis. A review recently has appeared of the interpretation of spectral parameters.[2]

Extraction of NMR Parameters

Structural information of peptide conformation is obtained from interpretation of such NMR parameters as chemical shift (δ), scalar coupling (J), nuclear Overhauser enhancement (NOE), and relaxation times (T_1, T_2, $T_{1\rho}$). Spectral analysis and often special pulse techniques and measuring conditions are necessary to obtain these parameters. Recent advances in instrumentation and the development of new techniques allow reliable and careful extraction of the parameters even for relatively complex molecules.

Chemical Shift

Carbon-13 chemical shift values are readily obtainable from a proton broadband-decoupled spectrum. However, in proton spectra, because of homonuclear coupling, broad signals are obtained and more or less extensive overlap of signals occurs, which hinders the determination of proton chemical shift values. In those cases in which the overlap is less extensive or where only some signals are of interest this problem can be solved by difference decoupling.[22] This method usually is used for investigations of temperature and solvent dependence of NH chemical shifts, when NH signals are covered by signals of aromatic protons or the solvent. The difference spectrum is gained by substracting the spectrum without decoupling (decoupler frequency off resonance) from a spectrum in which the coupling α proton is irradiated selectively. The advantage of this technique is the elimination of those signals that are not coupled to the irradiated proton and that may cover the signals of interest.

A more general solution is to use a projection of a two-dimensional (2D) J, δ spectrum.[23] The 2D J, δ spectrum gives information about the chemical shift values in one dimension and about the coupling constants in the other. By projecting the two-dimensional spectrum a "broadband-decoupled" proton spectrum is obtained that can be used to determine chemical shift values. This is shown in Figure 6-2 for the cyclic tripeptide cyclo[-L-Pro-L-Pro-D-Pro-].[24,25] It is obviously impossible to get the chemical shift values of the protons β_1^c, γ_1^t, and β_2^t, which resonate at about 1.6 ppm, from the conventional 1D NMR

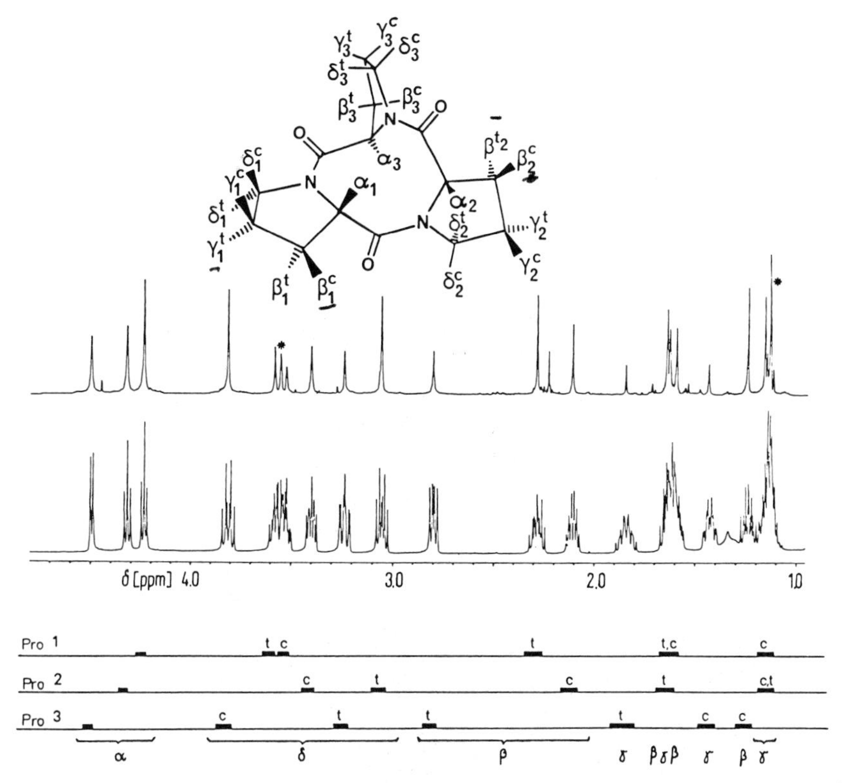

Figure 6-2. Projection of the 500-MHz J,δ spectrum of cyclo[-L-Pro$_2$-D-Pro-] in $CDCl_3$–C_6D_6 (1:8) (top) in comparison with the one-dimensional spectrum (middle). Assignments of the proline protons are given at the bottom (c = cis to α-H, t = trans to α-H). The peaks marked with an asterisk arise from strong AB effects.

spectrum. However, they are easy to obtain from the projection of the 2D J, δ spectrum. The chemical shift values obtained from the projection have been used as starting parameters for a simulation of the spin system. Sometimes there appear additional signals that result from strongly coupled spin systems. Those signals are marked with an asterisk in Figure 6-2. They possess a mean chemical shift between the signals of the two strongly coupled nuclei.[26]

Sometimes signals of interest are covered by solvent signals or other uncoupled singlet signals. Double quantum filtering can be applied to suppress such unwanted signals in 1D or 2D NMR spectra.[27-29] Figure 6-3 shows the stacked plot of the 2D J, δ spectra of cyclosporin A, a cyclic undecapeptide of great pharmalogic importance. The effect of the double quantum filter in the lower spectrum is obvious. One of the N-methyl group signals that covers other signals of interest is also removed and now the hidden signals can be evaluated.

Signals that are hidden in a group of other signals also can be recognized by their crosspeaks in two-dimensional homonuclear correlated spectra (correlated spectroscopy, or COSY, = Jeener spectroscopy; spin-echo correlated spectroscopy, or SECSY).[30-33] Running COSY spectra for studying temperature

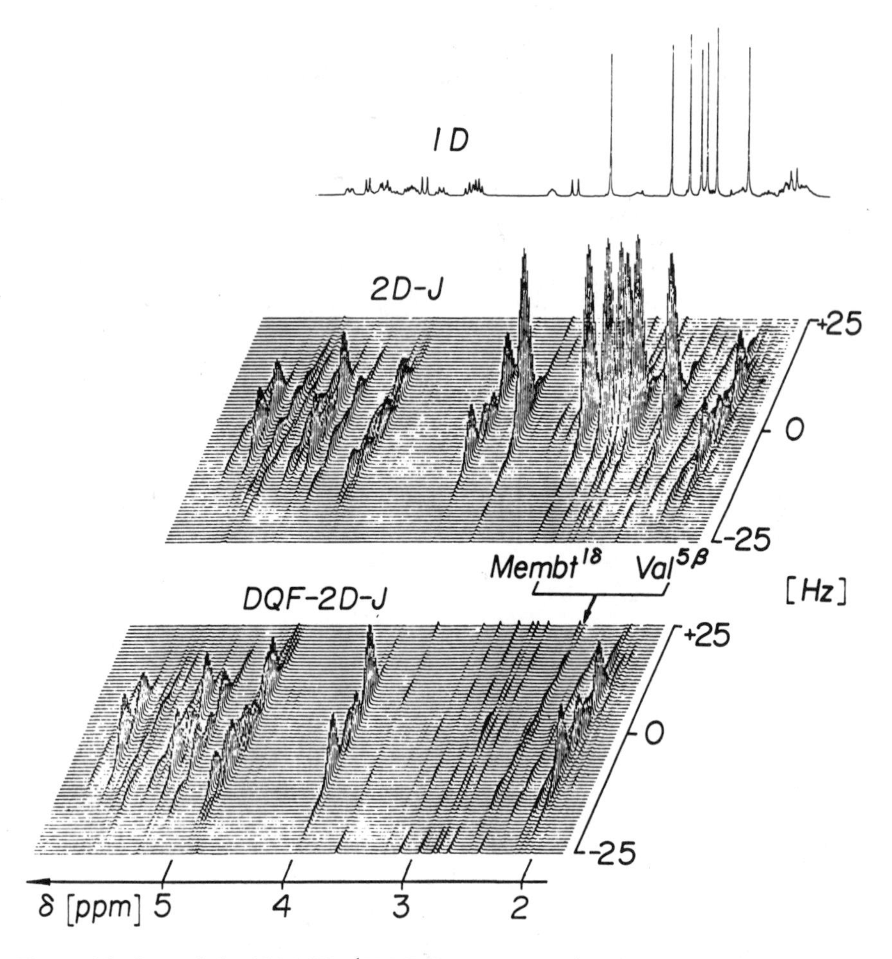

Figure 6-3. Part of the 300-MHz ^1H NMR spectrum of cyclosporin A in C_6D_6. Top: conventional one-dimensional spectrum between 2 and 6 ppm. Middle: stacked plot of the J,δ-resolved spectrum. Bottom: stacked plot of the double-quantum filtered J,δ spectrum. (Reprinted with permission from ref. 29).

and solvent dependence of NH chemical shift values is very time consuming and uneconomical, but in some cases such studies can help to solve the problem of signal identification. Such a procedure has been applied to cyclosporin A to unravel the strongly overlapping region, where 13 methyl groups resonate in a narrow chemical shift range of less than 0.5 ppm.[34]

Coupling Constants

Usually most of the ^1H, ^1H coupling constants are not directly obtainable from a one-dimensional NMR spectrum. Difference decoupling[22] allows a first estimation of coupling constants, but care should be taken because lineshapes may be distorted and splittings observed in the decoupling difference spectrum may be larger than the coupling constant. Another possibility is to obtain the coupling constants from a 2D J,δ spectrum. As mentioned above, a 2D J,δ spectrum presents information in one dimension about the chemical shift value (ω_2) and in the other about coupling constants (ω_1). By using cross sections of the chemical shift values it is possible to obtain the multiplet structure of each signal and then to determine coupling constants. The 2D J,δ spectra have the advantage of high resolution in the ω_1 dimension, because line-broadening effects caused by inhomogeneity are refocused, but this resolution is limited by the loss of magnetization because of relaxation during the evolution time t_1. Care should be taken in evaluating coupling constants because their values can be influenced by the data manipulation. The J,δ spectra of all common amino acids have been simulated and presented by Wider et al.[26] They also discuss the problem of strong coupling in J,δ spectra of amino acids.

The full homonuclear coupling information is also present in the crosspeaks of a normal H–H COSY or H–X COSY. New NMR instrumentation allows high resolution in 2D spectra, but because of line broadening resulting from data manipulation, it is difficult to evaluate conventional COSY spectra. These difficulties are now overcome by phase-sensitive COSY spectra.[35,36] We recently developed a convenient method to obtain coupling constants via differences and sums within phase-sensitive COSY spectra (DISCO).[98,99]

Heteronuclear coupling constants, except those of directly coupled nuclei, are usually difficult to obtain. Because they contain important stereochemical information via Karplus-type relationships,[37–39] it is of special interest that 2D techniques are now available that make it possible to get those values. In a 2D J,δ experiment, in which a selective 180° pulse to one proton is applied, all $^2J_{CH}$ and $^3J_{CH}$ coupling constants to that proton are obtained.[40]

This experiment was performed for all seven protons of the D-proline ring in cyclo[-L-Pro$_2$-D-Pro-].[41] This residue was chosen because all its protons are well-separated in $CDCl_3$–C_6D_6 (1:8), which is prerequisite of such an experiment. A cross section of the carbonyl signal of D-Pro in the 2D J,δ spectrum obtained by selective irradiation of the β_3^c proton is exhibited in Figure 6-4 in comparison to the fully coupled carbonyl signal. The eight $^3J_{CH}$ coupling constants along the pathway *C*–C–C–*H* are compatible with a Karplus curve evaluated from norbornane couplings[42] (Figure 6-5). The four values of $^3J_{CH}$ along *C*–N–C–*H* do not allow a proof of the two different Karplus equations proposed for such units.[37,40,43]

NOE Values

Nonequilibrium magnetization of a nucleus is transferred via dipole–dipole relaxation through space to neighboring nuclei. Intensity changes induced by these processes are called nuclear Overhauser enhancement effects (NOE effects).[22,44] Their importance for molecular conformations results from the dependence of NOE on nuclear distances.[45]

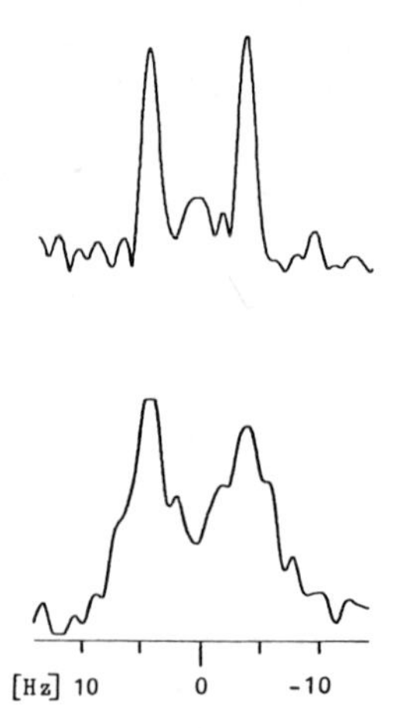

Figure 6-4. D-Proline-carbonyl signal of cyclo[-L-Pro_2-D-Pro-]. Above: cross section of a selective heteronuclear J,δ spectrum obtained by irradiating the β_3^c proton at 300 MHz. Below: fully coupled one-dimensional spectrum.

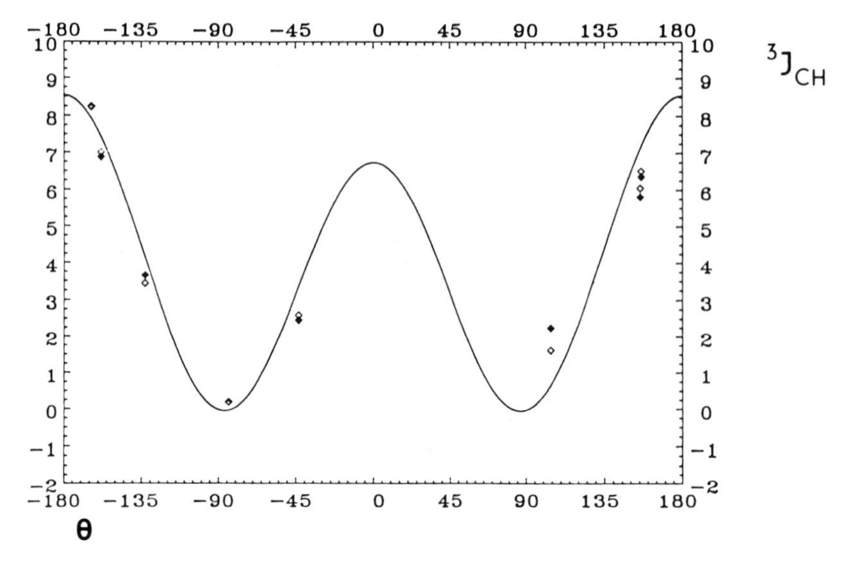

Figure 6-5. Karplus curve for the *H*–C–C–*C* fragment from the function[42] $^3J = 7.66 \cos^2\theta - 0.90 \cos\theta - 0.02$. Data points of D-Pro in cyclo[-L-Pro_2-D-Pro-] are obtained from phase-corrected spectra (◆) and from absolute value spectra (◇).

The buildup rate of NOE effects is time dependent and proportional to relaxation rates. A steady-state NOE effect is determined via NOE difference spectroscopy.[22,46] For each nucleus a separate difference spectrum is measured, which makes this technique very time-consuming in larger molecules.

A two-dimensional NOE spectrum (*N*uclear *O*verhauser and *E*xchange *S*pectroscop*y*, NOESY) allows a direct observation of all NOE effects in a molecule.[47-50] Especially for large molecules, when spin diffusion[51] distributes the desired NOE effect over longer distances into the molecule, NOESY is recommended.[49] Excellent examples of NOESY spectra and their combination with COSY spectra have been presented by Wüthrich's group for small proteins, such as BPTI[17,18,49,52] and glucagon.[18] It is possible to measure buildup rates of NOEs by the variation in the mixing time between the second and third 90° pulse.[49] Nuclear Overhauser enhancement effects in proteins are large and negative. With decreasing molecular size the NOE effects become more positive. Peptides of medium size, such as cyclohexapeptides, often show no effects at all. Hence, in these cases, the application of 2D NOE spectroscopy is more difficult because the detection of low-intensity crosspeaks is required.[53] Changing the measuring frequency, the solvent, and/or the temperature is often advised to obtain spectra of sufficient quality.

The high resolution of NOESY compared to NOE difference spectroscopy is a special advantage even for small molecules, such as cyclo[-L-Pro$_2$-D-Pro-]. This is demonstrated in Figure 6-6. Together with vicinal coupling constants, these effects have been used to identify the diastereotopic methylene protons in the proline ring system.[24]

Heteronuclear NOE measurements are important in proving the dipole–dipole relaxation mechanism for the interpretation of ^{13}C spin lattice relaxation times (T_1). The classical "gated Overhauser" experiment is performed for this purpose.[54]

A selective heteronuclear NOE effect from NH protons to carbonyl carbons can be used to determine intramolecular H–C distances. In such 1D NMR experiments one N*H* signal is irradiated selectively and NOE effects are measured at the fully coupled carbonyl carbon signals in the difference

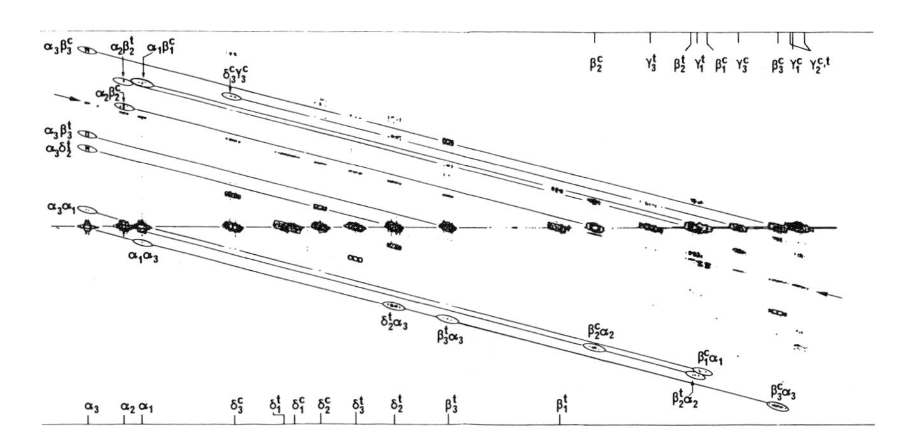

Figure 6-6. The 500-MHz NOESY spectrum (SECSY-type) of cyclo[-L-Pro$_2$-D-Pro-]. Crosspeaks that indicate NOE effects between two protons in different proline rings, as well as those that are important for assigning diastereotopic protons within the prolines, are circled.

technique.[55,56] Although it has been claimed that this technique can be used to identify NH···OC hydrogen bonds,[55] it should be pointed out that often more than one carbonyl group is close enough to produce similar NOE effects. Nevertheless, this technique may yield some information about molecular conformation that is not available otherwise.[56]

Spin–Lattice Relaxation Times

Spin–lattice relaxation times (T_1) are important for the determination of individual mobilities of backbone segments and side chains.[57] Their determination via classical one-dimensional methods[54] is well known and is not considered here.

Spectral Assignments

When correct NMR parameters have been obtained they must be assigned to a molecular constitution before their interpretation with respect to molecular conformation can start. In praxi, extraction and assignments proceed more or less simultaneously but it is strongly recommended not to begin conformational discussions before this work is finished! This discussion of assignment is divided into two parts. In the first, focus is on the identification of the individual amino acids and in the second on sequence analysis.

Recognition of Individual Amino Acids

Homonuclear Correlated Spectroscopy (H–H COSY and Variants). Most of the common amino acids exhibit typical patterns resulting from specific spin systems with distinct chemical shift values.[58] Identification of the connectivities is performed via homonuclear correlated spectroscopy (COSY) or SECSY.[30-33] In general the COSY-type spectrum is to be preferred because coupling across more than half the total spectral width usually occurs and maximal resolution is then higher than in a SECSY spectrum.[59] The expected crosspeaks in COSY spectra of different amino acids are shown schematically in Figure 6-7.

As can be seen in Figure 6-7, some of the amino acids show identical subspectral pattern; eg, the aromatic amino acids Phe, Tyr, Trp, and His all contain a AMXY spin system with similar chemical shifts and coupling constants. Methods for discriminating such systems are given below. Difficulties in interpreting a COSY spectrum often result from the ambiguities caused by signal overlap. There are various methods of overcoming this problem. When there is an extensive overlap of signals on the diagonal of the COSY spectrum, variation of the pulsewidth of the second pulse can be used.[30,33] An example is presented in Figure 6-8.

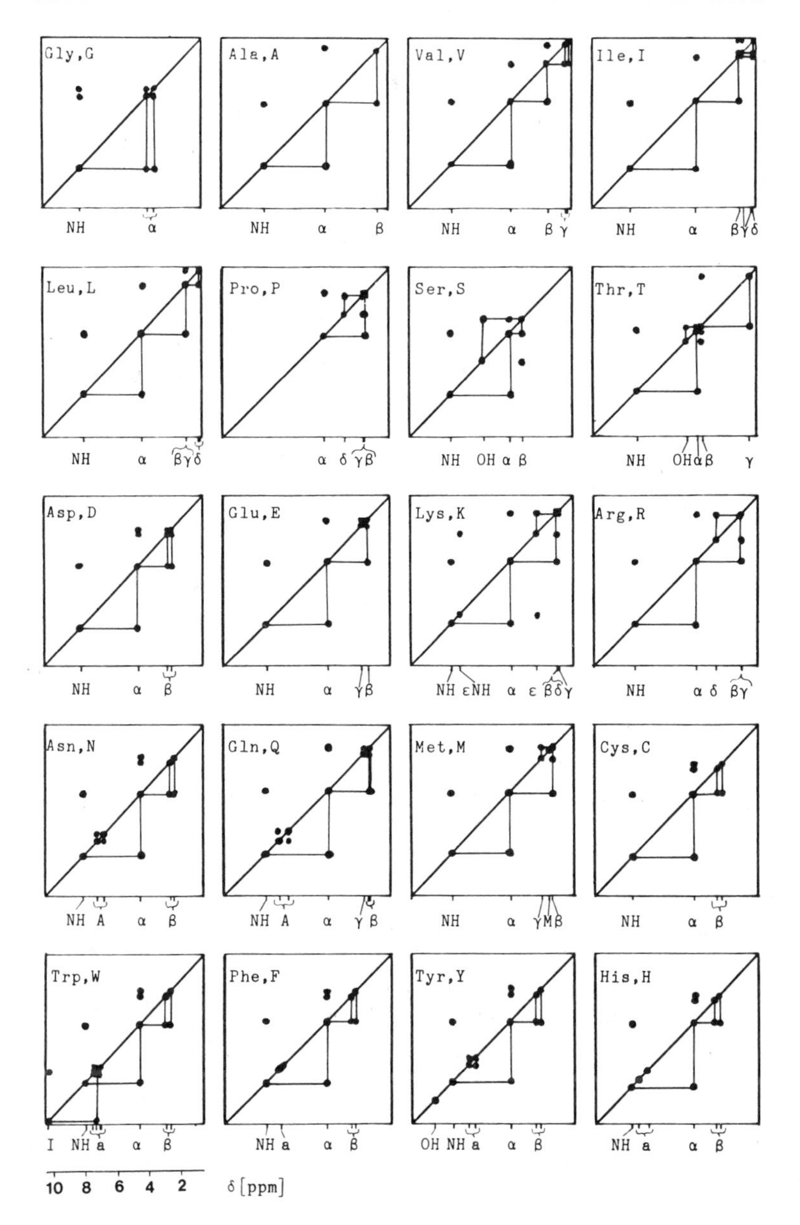

Figure 6-7. Schematic representation of H–H COSY spectra of the common amino acids from 0.5 to 10.5 ppm according to the scale given at the bottom of the figure. Abbreviations of the amino acids follow the IUPAC convention[9] (both the three- and the one-letter conventions are given). Most of the chemical shift values in DMSO are taken from ref. 58, p. 51. Of course, they depend on molecular constitution and conformation. Not included are signals that do not resonate in the given range; the COOH of Asp and Glu, the sidechain of His, and the Arg-NH and Cys-SH signals are not indicated. Additional abbreviations are for the following signals: A = amide protons, a = aromatic signals, M = *S*-methyl, I = indole NH.

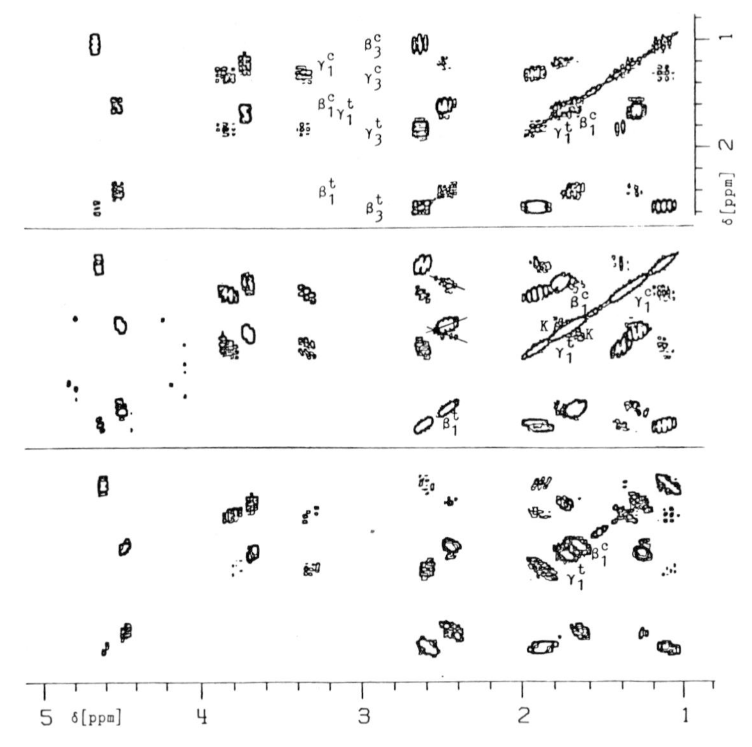

Figure 6-8. Part of the H–H COSY spectra of cyclo[-L-Pro-Bzl·Gly-D-Pro-] at 300 MHz in C_6D_6 with mixing pulse variation. Top, 90°; middle, 45°; bottom, 135°.

There are three spectra of cyclo[-L-Pro-Bzl-Gly-D-Pro-][60] with (a) 90°, (b) 45°, and (c) 135° pulses, respectively, as mixing pulse. In spectrum (a) it is impossible to recognize the crosspeaks between the signals of the γ_1^t and the β_1^c proton at 1.6–1.8 ppm. In spectrum (b) the diagonal is obviously narrower so that these crosspeaks become apparent. Besides, it is not possible to determine the chemical shifts of the two signals at the diagonal of the spectrum. These can be obtained easily by using the 135° pulse (spectrum c). In both spectra (b) and (c) the crosspeaks seem to be tilted. This effect may be utilized to evaluate the relative signs of the coupling constants from H–H COSY. For example, the orientation of the three crosspeaks of the β_1^t proton at 2.4 ppm in spectrum (b) can be used to identify the vicinal and geminal neighbors. Both crosspeaks to the γ protons exhibit the same inclination, ie, the same sign of the β,γ coupling constant, which is opposite to the geminal β,β coupling.

Overlap of crosspeaks cannot be removed by these methods. In this case a COSY spectrum with broadband decoupling in ω_1 can be utilized.[33,41]

When weaker signals are covered by strong singlet signals, such as solvent peaks or *N*-methyl signals, a COSY spectrum with a double-quantum filter (DQF) is recommended. Generally, phase-sensitive COSY combined with a

DQF is the best method to obtain spectra of highest resolution and small diagonal signals.[61]

When overlap of proton signals cannot be removed by the above-mentioned techniques, other homonuclear (relayed H–H COSY, multiple-quantum spectroscopy) or heteronuclear (H–C COSY, relayed H–C COSY, COrrelation via LOng range Coupling: H–C COLOC) methods can be applied. Before those are discussed however, it should be noted that often a change of solvent, temperature, and/or pH followed by 2D NMR techniques solves the problems arising from overlapping signals. This has been applied successfully to cyclosporin A in analyzing the very crowded region of the proton spectrum where the methyl groups resonate.[34,62] The advantage of using solvent titration followed by COSY spectra is that the crosspeaks are separated in the second dimension via the chemical shift of the coupling partners. It is unlikely that all protons responsible for the overlapping crosspeaks are shifted in the same way.

A frequent problem is that the coupling pattern within an amino acid cannot be followed along the whole sidechain for two reasons:

1. Signal intensity of a crosspeak can be too small to be detectable, for example when the proton is coupled to many partners, as with the γ proton of Leu, which is coupled to eight neighbors (Figure 6-9a).
2. When an amino acid is present twice or more in a molecule and the chemical shift of one signal of each amino acid overlaps. (Figure 6-9b).

One method for overcoming this problem is a relayed H–H COSY experiment in which the magnetization is transferred in two consecutive steps via proton spin-coupling (H_A–H_B–H_C).[36] Even if there is no direct coupling between protons A and C crosspeaks at the chemical shifts of A and C indicate their remote connectivity.

In principle, more than one magnetization transfer step can be performed to get further connectivities. One possible way to achieve this is to use additional transfer sequences in the relayed experiment. Another possibility is to run a *TO*tal *C*orrelation *S*pectroscop*y* (TOCSY), experiment, in which no adjustment to coupling constants needs be made and the advantage of a phase-sensitive spectrum is present.[63]

The TOCSY spectrum of cyclo[-L-Pro-Bzl·Gly-D-Pro-][60] is shown in Figure 6-10. Now each signal exhibits crosspeaks to each proton of the same proline system. Hence, the connectivities of the two seven-spin systems of the prolines are directly evident in such a spectrum. Generally, it is sufficient to have only one proton separated for each spin system to identify the other protons belonging to this system.

Aromatic amino acids all exhibit a similar AMXY spin pattern, which cannot be differentiated. However, their aromatic protons often show a typical subspectrum that may be used for their discrimination. In a normal spectrum the aromatic and the aliphatic parts seem to be totally separated but, in fact, there is a small coupling between aromatic and β protons. This can be used to identify the connectivity between both subspectra, when a COSY spectrum is

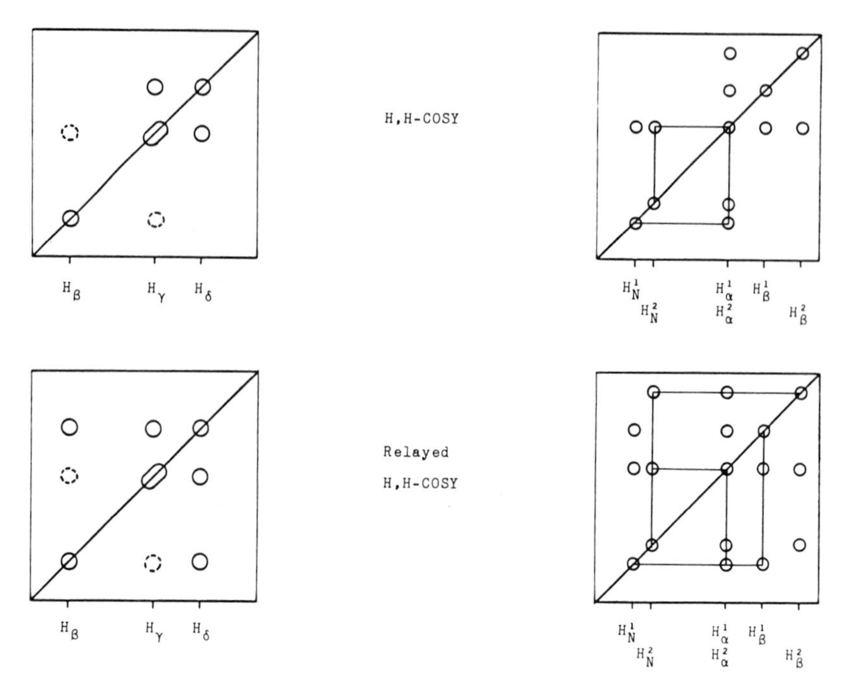

Figure 6-9. Schematic representation of H–H COSY (top) and relayed H–H COSY (bottom) spectra. The example on the left demonstrates the advantage of the relayed H–H COSY spectrum for a spin system in which the signal in the middle is very broad (eg, Leu γ proton). The example on the right shows two separate spin systems with overlapping central signals.

applied that is optimized for the detection of long-range couplings by delayed acquisition in the evolution time t_1 and the detection time t_2.[33] In such a spectrum the long-range coupling is indicated by crosspeaks. In evaluating this spectrum it must be remembered that strong coupling often yields crosspeaks of lower intensity than those in the normal COSY spectrum. An example is shown in the cyclic somatostatin analog cyclo[-MePhe-Phe-D-Trp-Lys-Thr-Phe-] in Figure 6-11, where the Trp residue is identified via the coupling from indolic NH to the aromatic neighbor in the pyrrole ring, which couples to both β protons at 2.66 and 2.88 ppm.[64] Another way of identification of Trp by H–C COSY is described in the section, Heteronuclear Correlation Spectroscopy. A number of other long-range couplings in benzylic systems from the three Phe residues (marked in Figure 6-11 by a, b, and c) and the protecting groups Z (Lys, d) and Bzl (Thr, e) are present. Tyrosine residues also are easy to identify because the typical AA′BB′ aromatic spin system shows long-range couplings of the BB′ protons to the phenolic O*H* as well as to the benzylic β-protons.[65]

N-Alkylated amino acids also exhibit long-range couplings between the *N*-alkyl group and the α protons. So 2D homonuclear correlated spectroscopy has been used to assign *N*-benzyl[60,66] or *N*-methyl signals[34] in cyclic tri-peptides and cyclosporin A, respectively.

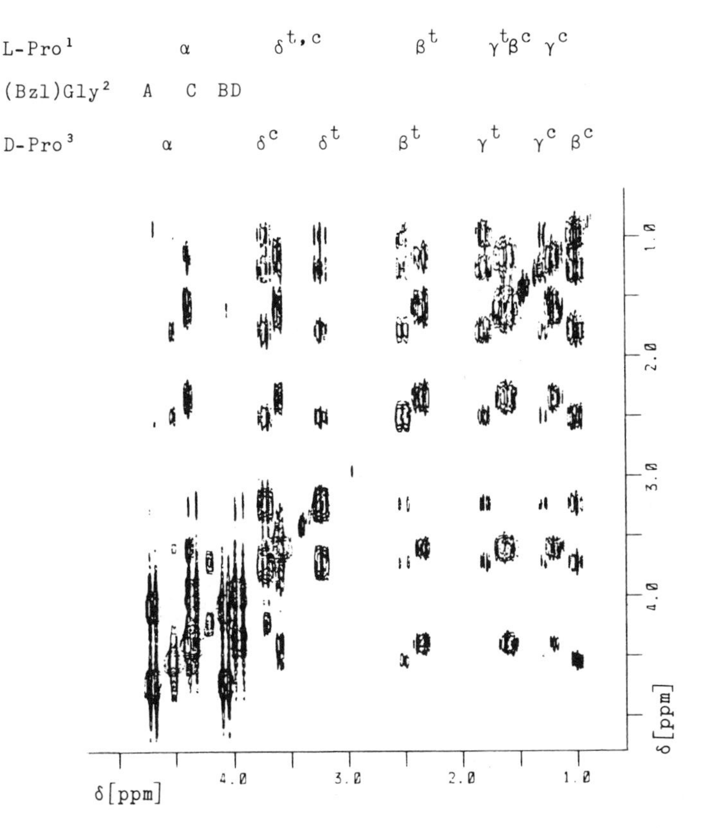

Figure 6-10. Part of the 270-MHz TOCSY spectrum of cyclo[-L-Pro-Bzl·Gly-D-Pro-] in C_6D_6 (compare Figure 6-8). The complete proline proton spectra are represented clearly at the β_1^t and β_3^t protons in the horizontal or the vertical axis. Some crosspeaks of the α protons are relatively small. The assignment is given at the top.

Multiple Quantum Spectroscopy. One-dimensional NMR spectroscopy and all 2D techniques described so far in this chapter only detect single-quantum coherences. It has been shown that 2D methods can be used to detect multiple-quantum coherences evolving during the t_1 dimension.[30,67–70]

The evaluation of a 2D double-quantum spectrum (DQS) is easy to perform. Whereas the ω_2 axis contains the normal proton spectrum, in the ω_1 axis coupled pairs of protons are spread according to the sum of their resonance frequencies. The DQ-spectrum of the above-mentioned cyclotripeptide cyclo[-L-Pro-Bzl·Gly-D-Pro-] (see Figure 6-10) is shown in Figure 6-12. All expected pairs of coupled protons are detected by optimizing experimental parameters to an average coupling constant of 20 Hz.[71] For the evaluation of such a spectrum it is important to note that all pairs appear symmetric to a line from the lower left side to the upper right side. For example, the horizontal bars 7 and 8 indicate the coupling of the α proton of the L-Pro residue to both β protons. Sometimes additional peaks (type II)[69] of remote connectivities appear. For

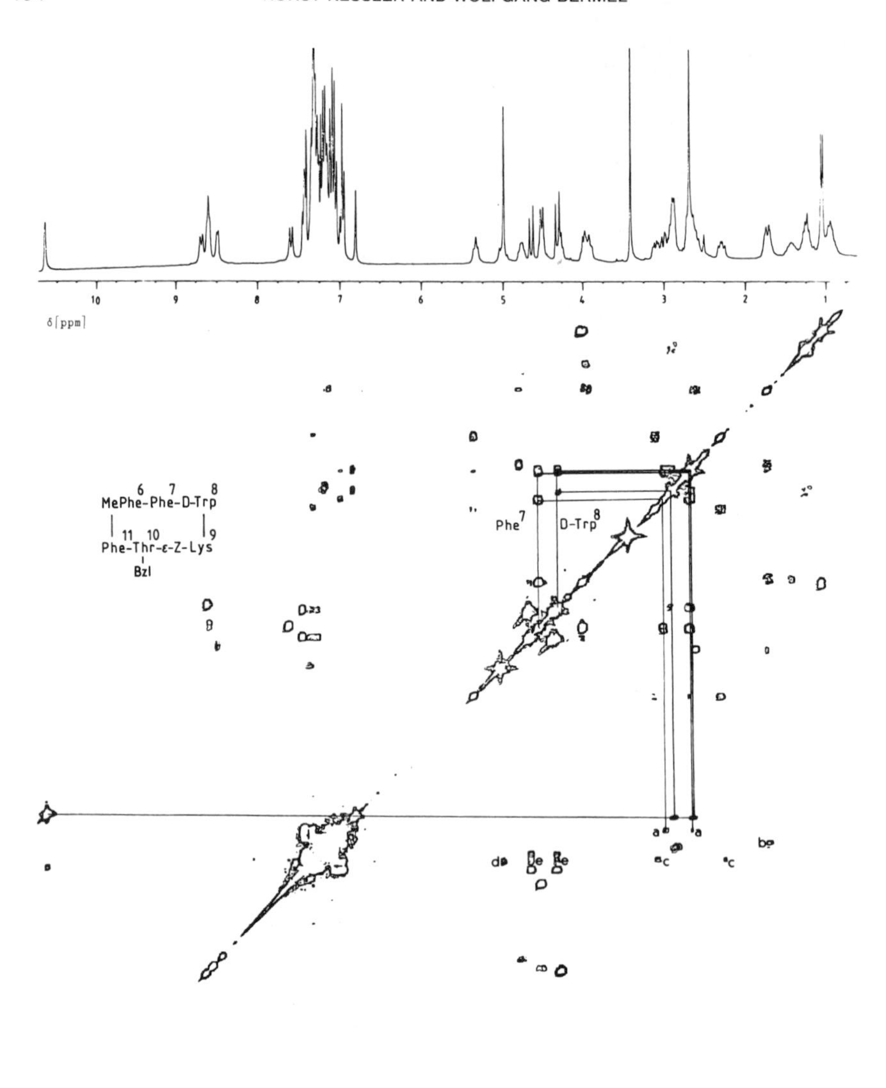

Figure 6-11. The 300-MHz H–H COSY spectrum, optimized for the detection of small couplings (delay: 100 ms each), of a cyclic hexapeptide. The connectivity within the Trp residue is shown by connecting lines. Crosspeaks indicating further long-range couplings: (a) arom-Phe^7-β; (b) arom-Phe^{11}-β; (c) arom-MePhe^6-β; (d) Z-protecting group of Lys; (e) Bzl· protecting group of Thr; (f) indole-NH-arom.

example, bar 16, caused by the coupling β^c–β^t, also shows a peak at the chemical shift of the α proton in the same proline ring (L-Pro).

Further applications to peptides are given in references 34, 72 and 73.

Heteronuclear Correlation Spectroscopy. The assignment of the spectra of

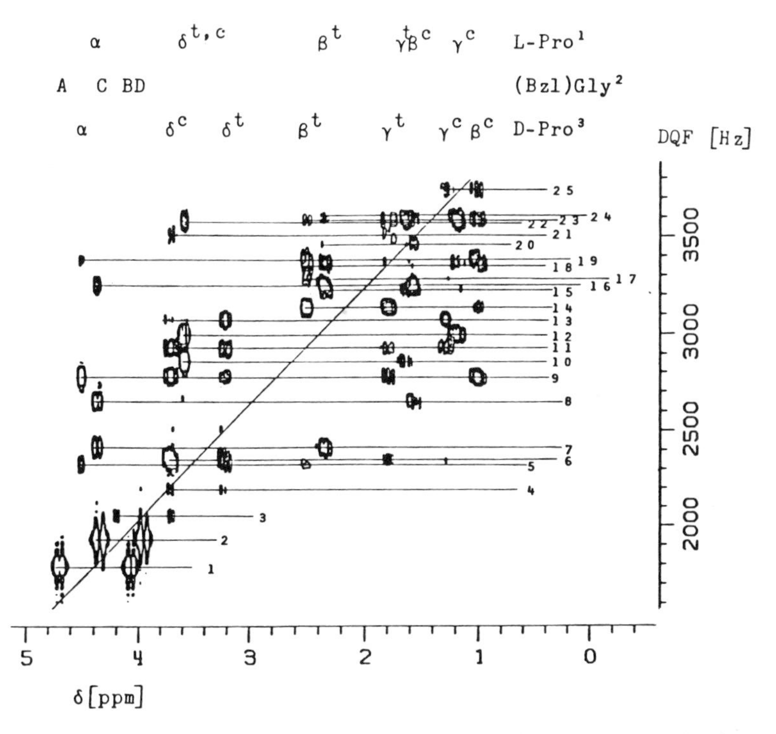

Figure 6-12. The 300-MHz double–quantum spectrum of cyclo[-L-Pro-Bzl·Gly-D-Pro-] in C_6D_6 (compare Figures 6-8 and 6-10). Assignment is given at the top.

heteronuclei, such as carbon[74,75] or nitrogen,[76] usually is done by a shift-correlation experiment (H–X COSY) based on the assigned proton spectrum.[6,60,76–79] In this experiment proton chemical shift values are correlated with those of the carbon and nitrogen atoms directly attached to the protons. An example is presented in Figure 6-13. Many of the carbon signals can be assigned from this spectrum, as is discussed in detail in the literature.[79] A complete assignment of all aliphatic carbon atoms was possible only by applying relayed H–C COSY (see below). The characteristic highfield shift of the β carbon signal of Trp caused by the enamine structure in the indole ring can be used to identify this amino acid. Hence, it is often possible to discriminate the proton signals of the amino acids Phe and Trp via H–C COSY.[77,79]

There are some modifications of this method for ^{13}C spectra. One of these improves the signal to noise ratio by eliminating proton–proton coupling.[80] This is referred to as proton decoupling in ω_1.

Another experiment, H–C COLOC,[81,82] uses proton–carbon long-range couplings instead of one-bond CH couplings for the assignment of ^{13}C signals and also employs a broadband decoupling in ω_1. The H–C COLOC spectrum yields connectivity information of the carbon atoms, as demonstrated in Figure 6-14 for cyclo[-L-Pro_2-D-Pro-]. Heteronuclear coupling to many protons at each

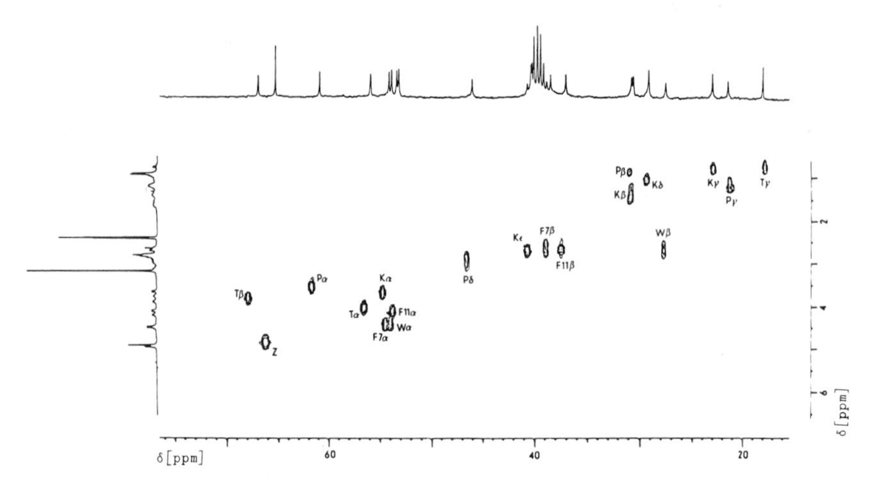

Figure 6-13. Aliphatic range of the 300-MHz H–C COSY of cyclo[-Pro-Phe-D-Trp-Lys-Thr-Phe-] in DMSO.

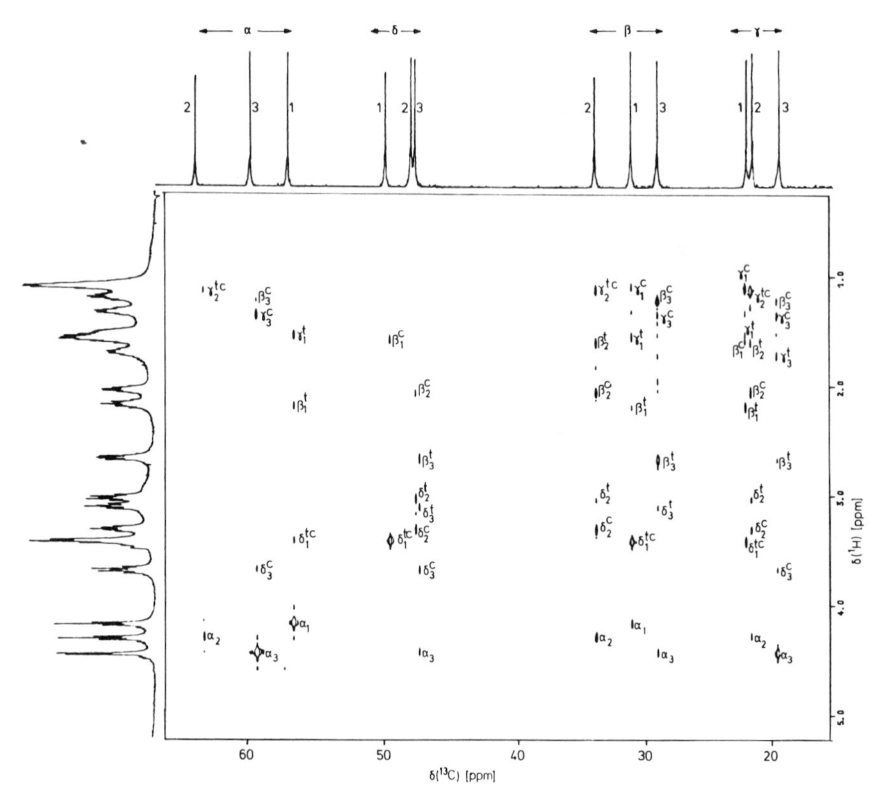

Figure 6-14. The H–C COLOC spectrum of the aliphatic carbons of cyclo[-L-Pro_2-D-Pro-] in $CDCl_3$–C_6D_6 (1:9) (see Figures 6-2, 6-4, and 6-6).

aliphatic carbon is observed. As an example the β carbon of the proline system 2 at 34.1 ppm exhibits crosspeaks for the protons α, δ_2^c, δ_2^t, β_2^c, β_2^t, and $\gamma_2^{c,t}$. A similar pattern is found at γ carbon-2 at 22.8 ppm, indicating the carbon connectivity β_2–γ_2. It is easy to see that for each carbon there is at least one crosspeak caused by long-range coupling across two or three bonds in addition to that of the directly attached CH pair. The complete carbon connectivities follow from an analysis of such a spectrum.[83] If necessary, it is possible to suppress the crosspeaks from one-bond coupling by a low-pass J filter.[84]

A special feature of H–C COLOC results from the fact that quaternary carbons also exhibit crosspeaks. Therefore, carbonyl carbons can be assigned via crosspeaks to neighboring amide protons, and α and β protons. This can be used for peptide sequencing and is discussed in the section, Sequence Analysis.

Carbon connectivity information is also present in the relayed H–C COSY spectrum.[79,85,86] In a relayed carbon–proton shift correlation, at each carbon chemical shift, in addition to the peak for the directly attached proton, signals for all other protons that couple to it also are shown. This technique is especially useful when proton signals overlap and conventional H–C COSY allows no discrimination. This was the case in the somatostatin analog cyclo-[-Pro^6-Phe^7-D-Trp^8-Lys^9-Thr^{10}-Phe^{11}-], where the α protons of Phe^7 and D-Trp^8 coincide and severe overlap occurs in the β-region. The relayed peaks to the NH protons at the α carbons (coupling pathway: H_N–H_α–C_α) as well as those

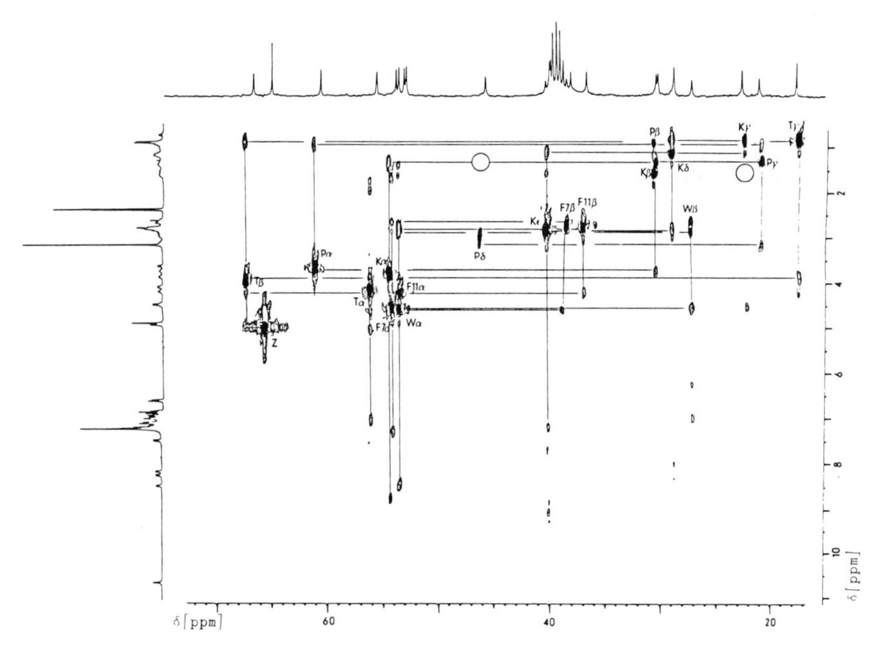

Figure 6-15. Relayed H–C COSY spectrum of cyclo[-Pro-Phe-D-Trp-Lys-Thr-Phe-] in DMSO (see Figure 6-13). The circles indicate missing relayed peaks.

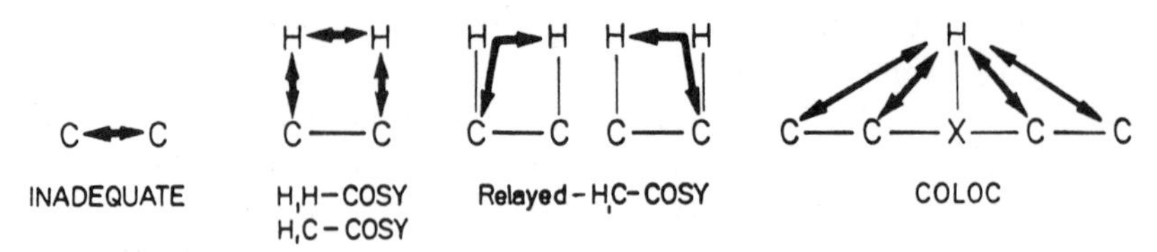

Figure 6-16. Carbon connectivities via homo- and heteronuclear couplings used in different NMR techniques.

from the α protons to the β carbons (H_α–H_β–C_β) lead to the complete carbon assignment (Figure 6-15).[79]

A further advantage of relayed H–C COSY is that protons attached to nitrogen (amide) and oxygen (HO, when coupling to the neighbored proton can be detected, eg, in dimethylsulfoxide solution[87]) can be observed indirectly via crosspeaks at the carbon.

Carbon connectivities can be detected directly in a 2D INADEQUATE spectrum, which uses double-quantum spectroscopy of ^{13}C–^{13}C pairs.[88,89] However, this experiment is quite insensitive because only those molecules containing two ^{13}C isotopes attached to each other contribute to the signal. A large amount of substance with high solubility is required. Therefore its application to peptides is severely restricted.

The different possible ways of elucidating carbon connectivities are summarized in Figure 6-16. The most common procedure for this purpose is the combination of H–H COSY and H–C COSY. If overlapping proton signals prevent a straightforward assignment, the use of relayed H–C COSY or H–C COLOC is recommended. Both experiments are less sensitive than a H–C COSY, however, and usually are performed in overnight runs with 150 mg of a peptide, having molecular weight of 1000 at 300 MHz. In contrast, H–C COSY can be performed with 50–100 mg under the same conditions.

Sequence Analysis

The amino acids of a peptide can be treated in first approximation as isolated spin systems separated by the amide bonds. For sequence analysis, through-space (NOE) or through-bond (coupling) effects across the amide bonds must be used.

The conventional way to get the connectivity between the amino acids is to measure NOE effects. The most common effects are observed between NH protons of one amino acid and the α or β protons of the amino acid preceding the sequence.[2] Nuclear Overhauser enhancements between α protons of adjacent amino acids occur across *cis*-peptide bonds.[24,56,79] Nuclear Overhauser enhancement effects are detected either by one-dimensional NOE difference spectroscopy or by NOESY (see section NOE-Values).

Although proton spectra of peptides contain apparently separated amino acid spin systems, small proton–proton couplings, as well as two- and three-

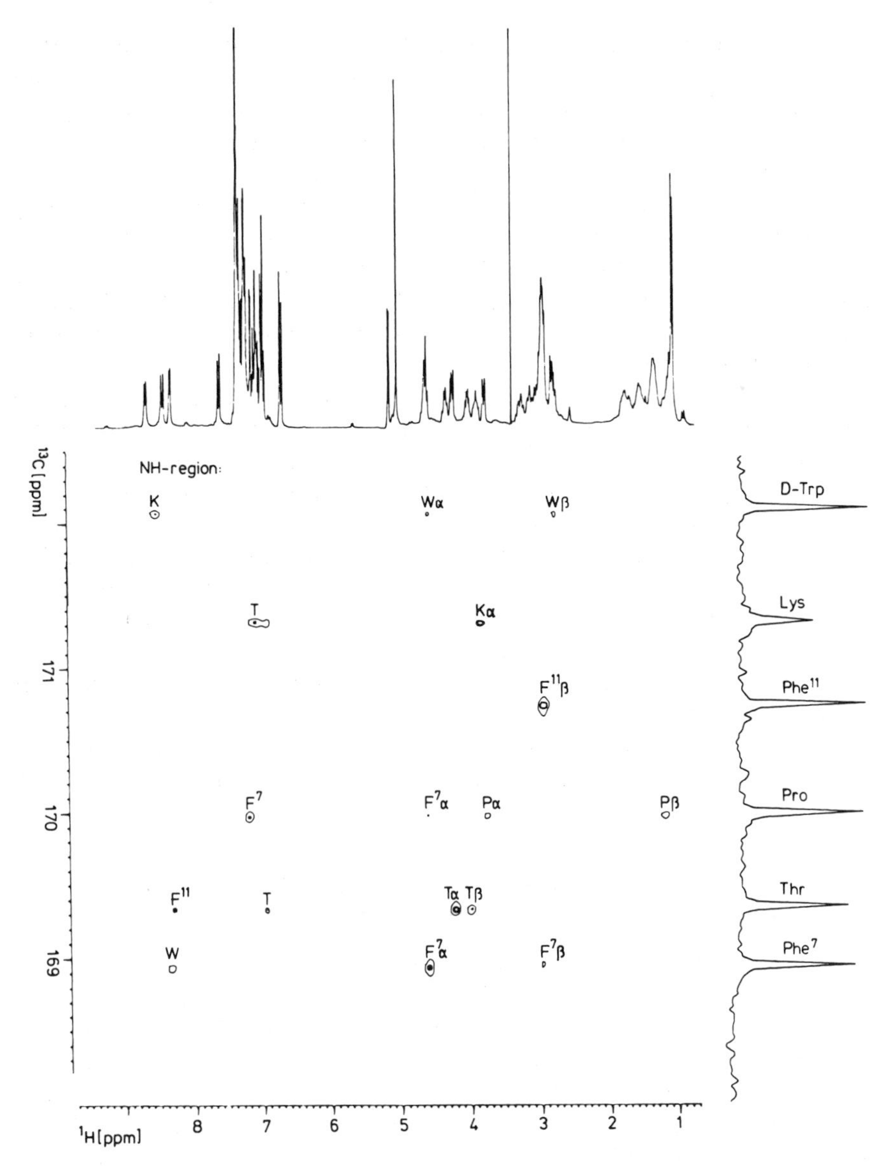

Figure 6-17. The H–C COLOC spectrum of the carbonyl groups of cyclo[-Pro-Phe-D-Trp-Lys-Thr-Phe-] in DMSO (see Figures 6-13 and 6-15).

bond coupling of carbon and nitrogen atoms to protons of the neighboring amino acids, are present.

Proton coupling can occur between the α proton of one amino acid and the α and NH proton of the next amino acid. To our knowledge, H_{α}^{i}–H_{N}^{i+1} coupling has not yet been observed. Coupling between α protons usually is found across *cis*-peptide bonds[24,60,90,91] but see also ref 100. Small but significant couplings exist also in peptides that contain *N*-alkylated amino acids, including proline. Those couplings between the protons in the α position of the *N*-alkyl group and the α protons of the same and the preceding amino acid can be used for sequencing.[24,34,60] None of these long-range couplings can be detected in the one-dimensional spectrum. Their identification is performed via a H–H COSY optimized for this purpose.[33] Recently, coupling from carbonyl carbons to the NH proton of the following amino acid has been used for peptide sequencing.[34,81,92,93] Further coupling of the carbonyl carbon can occur to the α and β proton within the same amino acid, as well as to the α proton of the following one. Unequivocal assignment of the carbonyl group is provided when coupling to a β proton is observed.

In our experience the coupling to the α proton within the same amino acid is always stronger than that to the α proton of the residue following in the sequence. This is demonstrated in Figure 6-17 for the cyclic somatostatin analog cyclo[-Pro-Phe-D-Trp-Lys-Thr-Phe-], which has been discussed in the section on Heteronuclear Correlation Spectroscopy. The H–C COLOC spectrum[81,82] exhibits crosspeaks for each of the six carbonyl carbons. For example, the Pro-CO is assigned via its coupling to the Pro-β proton and contains the sequence information in the coupling to the Phe-7 NH and the α proton. Following the sequence, the Phe-7-CO group exhibits a coupling to the Trp NH. The sequence therefore is Pro-Phe^7-Trp, which can be extended easily to the full sequence.

It is evident that when the sequence is already known, the H–C COLOC-technique can be used to identify the amino acids in the spectrum as well.

Dynamic Phenomena

Nuclear magnetic resonance spectroscopy is also an appropriate tool for following slow chemical exchange processes with free activation enthalpies $\Delta G^{\ddagger}$ between 5 and 25 kcal mol^{-1} at 298 K. An example of this is the rotation about *cis-trans* Xxx-Pro peptide bonds. The identification of exchanging nuclei is performed via NOESY.[46-49,94,95] Crosspeaks in a NOESY spectrum are now caused by magnetization transfer via chemical exchange instead of dipolar relaxation (NOE effects). When NOE effects are negative, it is not possible to determine a crosspeak's origin from its appearance, but temperature effects may be used to discriminate NOE and exchange effects. In case of positive NOE effects a phase-sensitive NOESY[101] enables their discrimination from exchange peaks due to the different sign. In peptides, 2D exchange spectroscopy is especially helpful in identifying mutually exchanging signals.[66,96] This is seen in Figure 6-18 for a cyclic enkephaline analog[96] that exists as a mixture of two conformations in the ratio 40:60 at room temperature in DMSO solution. Exchange is observed by the crosspeaks above and below the horizontal line ($\omega_1 = 0$) in the SECSY-type NOESY spectrum. For chemical exchange a SECSY-type representation is

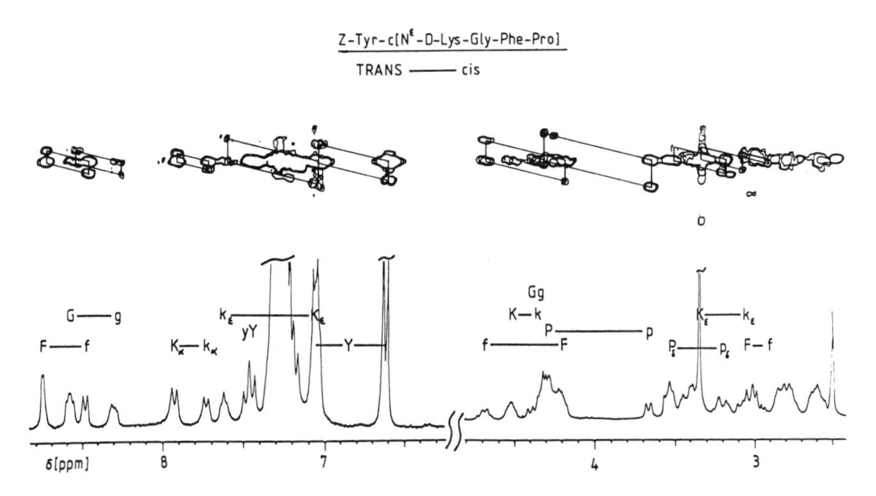

Figure 6-18. The NOESY spectrum (SECSY type) of a cyclic enkephaline analog. Small capital letters in the one-letter notation of amino acids are used to indicate chemical exchange of signals of the trans (cis) conformation in the one-dimensional spectrum below.

preferred over a COSY type because the difference in chemical shift between two signals belonging together in conformational isomers is usually small (< 2 ppm in ^{1}H NMR) and resolution is then better in SECSY-type spectra. Exchange spectroscopy is also applicable to ^{13}C nuclei.[95,96]

The quantitative analysis is done with one-dimensional techniques provided that the signals of the exchanging nuclei are separated in the spectrum. Saturation transfer experiments are used in combination with selective and nonselective T_1 measurements.[97–100] The information gained includes the equilibrium constant K, the exchange rates, and the free activation enthalpy $\Delta G^{\ddagger}$. The exchange rates and the free activation enthalpy can also be determined by lineshape analysis.[10,11]

Sidechain mobility can be analyzed by spin–lattice relaxation times. Numerous examples can be found in the literature. As mentioned in the section, Spin-Lattice Relaxation Times, this point is not discussed further here.

Conclusions

The development of 2D NMR techniques is still continuing. The manifold of variations already available are impressive. By a combination of these various techniques, resolution enhancement and connectivity information that could not be gained before by one-dimensional techniques is now obtained quite easily. Different molecules demand different strategies for assignment and extraction of required NMR parameters. Economy, as well as specific restrictions (solubility, solvent, relaxation times, amount of sample, available measuring frequency, instrumentation), require selecting experiments carefully to solve a problem. Hence, it is difficult to recommend a general strategy that holds for every peptide.

Nevertheless, it is useful to start with 1D spectra to determine the best solvent and concentration conditions. A homonuclear COSY spectrum is then performed. The best version is the phase-sensitive mode with a double-quantum filter (to filter out dispersive contributions on the diagonal). The H–C COSY technique often enables the assignment of many (or all) carbon signals. After that, a decision must be made about further experiments, such as homo- or heteronuclear relayed COSY or H–C COLOC, for specific spin system analysis or sequencing, respectively. Also a COSY with delay, to identify connectivities via proton long-range couplings, could be helpful. State of the art NMR spectroscopy in many cases now allows an unequivocal assignment of all NMR signals to the constitution and the extraction of spectral parameters. This is the indispensible prerequisite for a reliable discussion of the conformation.

The accuracy of the statements about molecular conformation strongly depends upon the size of the molecule. For example, a very detailed picture of cyclic tripeptides with regard to the static and dynamic behavior in crystal and solution is now available. Even spectra of molecules of molecular weight of more than 1000, such as cyclosporin A, can now be assigned completely. However, it is still not yet possible to obtain the same level of confidence for medium-sized molecules. In addition complete assignments of all magnetic active nuclei in small proteins is nearly impossible and, in many cases for very large molecules, only a few of the resonances can be assigned. Nevertheless, there can be no doubt that the limits of handling molecules are greatly enlarged by the introduction of the 2D NMR techniques.

Acknowledgments

We thank the Deutsche Forschungsgemeinschaft and the Fonds der Chemischen Industrie for financial support of this work. W.B. thanks the Fonds der Chemischen Industrie also for a personal grant. We kindly acknowledge discussions and assistance from our co-workers I. Damm, C. Griesinger, Dr. G. Hölzemann, H. Kogler, H. Oschkinat, and J. Zarbock, as well as Dr. W. E. Hull, Bruker, Karlsruhe. Some peptide samples were obtained from our co-workers Dr. M. Bernd and Dr. G. Krack as well as from SANDOZ AG.

References

1. Kessler, H.; Hölzemann, G.; Zechel, C. *Int. J. Pept. Prot. Res.* **1985**, *25*, 267.
2. Kessler, H. *Angew. Chem.* **1982**, *94*, 509; *Angew. Chem. Int. Ed. Engl.* **1982**, *21*, 512.
3. Hruby, V.; Hadley, M. E. "Binding and Information Transfer in Conformationally Restricted Peptides." Paper presented at the 18th Solvay Conference in Chemistry, Nov. 28–Dec. 1, 1983, Brussels, Belgium.

4. Kessler, H. "Designing Activity and Receptor Selectivity in Cyclic Peptide Hormone Analogues." Paper presented at the 18th Solvay Conference in Chemistry, Nov. 28–Dec. 1, 1983, Brussels, Belgium.
5. Kessler, H.; Zimmermann, G.; Förster, H.; Engel, J.; Oepen, G.; Sheldrick, W. S. *Angew. Chem.* **1981**, *93*, 1085; *Angew. Chem. Int. Ed. Engl.* **1981**, *20*, 1053.
6. Kessler, H.; Schuck, R.; Siegmeier, R.; Bats, J. W.; Fuess, H.; Förster, H. *Justus Liebigs Ann. Chem.* **1983**, 231.
7. Karle, I. L. in "The Peptides", Gross, E.; Meienhofer, J., Eds.; Academic: New York, 1981; Vol. 4, p. 1; Snyder, J. P. *J. Am. Chem. Soc.* **1984**, *106*, 2393.
8. Madison, V.; Adreyi, M.; Deber, C. M.; Blout, E. R. *J. Am. Chem. Soc.* **1974**, *96*, 6725.
9. IUPAC-IUB Commission on Biochemical Nomenclature. *J. Mol. Biol.* **1970**, *52*, 1; *Biochemistry* **1970**, *9*, 3471.
10. Kessler, H. *Angew. Chem.* **1970**, *82*, 237; *Angew. Chem. Int. Ed. Engl.* **1970**, *9*, 219.
11. Binsch, G.; Kessler, H. *Angew. Chem.* **1980**, *92*, 445; *Angew. Chem. Int. Ed. Engl.* **1980**, *19*, 411.
12. Jackman, L. M.; Cotton, F. A., Eds. "Dynamic Nuclear Magnetic Resonance Spectroscopy", Academic: New York, 1975; and references cited therein.
13. Kessler, H.; Rieker, A. *Justus Liebigs Ann. Chem.* **1967**, *708*, 57.
14. Jardetzky, O. *Biochim. Biophys. Acta* **1980**, *621*, 227.
15. Ramachandran, G. N.; Sasisekharan, V. *Adv. Protein Chem.* **1968**, *23*, 283.
16. Braun, W.; Bösch, C.; Brown, L. R.; Gō, N.; Wüthrich, K. *Biochim. Biophys. Acta* **1981**, *667*, 377.
17. Wüthrich, K.; Wider, G.; Wagner, G.; Braun, W. *J. Mol. Biol.* **1982**, *155*, 311.
18. Wagner, G.; Wüthrich, K. *J. Mol. Biol.* **1982**, *155*, 347.
19. Blout, E. R. *Biopolymers* **1981**, *20*, 1901.
20. Hruby, V. *Life Sci.* **1982**, *31*, 189.
21. Kessler, H.; Hölzemann, G.; Geiger, R. In "Proceedings of the 8th American Peptide Symposium", Hruby, V.; Rich, D. H., Eds.; Pierce Chemical Company: Rockford, Ill., 1983; p. 295.
22. Sanders, J. K. M.; Mersh, J. D. *Progr. Nucl. Magn. Reson. Spectrosc.* **1982**, *15*, 351.
23. Nagayama, K.; Bachmann, P.; Wüthrich, K.; Ernst, R. R. *J. Magn. Reson.* **1978**, *31*, 133.
24. Kessler, H.; Bermel, W.; Friedrich, A.; Krack, G.; Hull, W. E. *J. Am. Chem. Soc.* **1982**, *104*, 6297.
25. De Leeuw, F. A. A. M.; Altona, C.; Kessler, H.; Bermel, W.; Friedrich, A.; Krack, G.; Hull, W. E. *J. Am. Chem. Soc.* **1983**, *105*, 2237.
26. Wider, G.; Baumann, R.; Nagayama, K.; Ernst, R. R.; Wüthrich, K. *J. Magn. Reson.* **1981**, *42*, 73.
27. Piantini, U.; Sørensen, O. W.; Ernst, R. R. *J. Am. Chem. Soc.* **1982**, *107*, 6800.
28. Shaka, A. J.; Freeman, R. *J. Magn. Reson.* **1983**, *51*, 169.
29. Kessler, H.; Oschkinat, H.; Sørensen, O. W.; Kogler, H.; Ernst, R. R. *J. Magn. Reson.* **1983**, *55*, 329.
30. Aue, W. P.; Bartholdi, E.; Ernst, R. R. *J. Chem. Phys.* **1976**, *64*, 2229.
31. Nagayama, K.; Kumar, A.; Wüthrich, K.; Ernst, R. R. *J. Magn. Reson.* **1980**, *40*, 321.
32. Bax, A.; Freeman, R.; Morris, G. *J. Magn. Reson.* **1981**, *42*, 164.
33. Bax, A.; Freeman, R. *J. Magn. Reson.* **1981**, *44*, 542.
34. Kessler, H.; Loosli, H. R.; Oschkinat, H. *Helv. Chim. Acta* **1985**, *68*, 661.
35. Marion, D.; Wüthrich, K. *Biochem. Biophys. Res. Commun.* **1983**, *113*, 967.
36. Wagner, G. *J. Magn. Reson.* **1983**, *55*, 151.
37. Bystrov, V. F. *Progr. Nucl. Magn. Reson. Spectrosc.* **1976**, *10*, 41.
38. Hansen, P. E. *Progr. Nucl. Magn. Reson. Spectrosc.* **1981**, *14*, 175.
39. Marshall, J. L. "Carbon-Carbon and Carbon-Proton NMR Couplings"; Verlag Chemie: Deerfield Beach, Florida, 1983.
40. Bax, A.; Freeman, R. *J. Am. Chem. Soc.* **1982**, *104*, 1099.
41. Bermel, W. Dissertation, J. W. Goethe-Universität, Frankfurt, 1983.
42. Aydin, R.; Loux, J. P.; Günther, H. *Angew. Chem.* **1982**, *94*, 451.
43. Bystrov, V. F.; Gavrilov, Y. G.; Solkan, V. N. *J. Magn. Reson.* **1975**, *19*, 123.
44. Noggle, J. H.; Shirmer, R. E. "The Nuclear Overhauser Effect"; Academic: New York, 1971.
45. Billeter, M.; Braun, W.; Wüthrich, K. *J. Mol. Biol.* **1982**, *155*, 321.
46. Gibbons, W. A.; Crepaux, D.; Delayre, J.; Dunand, J.-J.; Wyssbrod, H. R. In "Peptides", Walter, R.; Meienhofer, J., Eds.; Ann Arbor Science Publ.: Ann Arbor, Mich., 1975; p. 127.

47. Macura, S.; Huang, Y.; Suter, D.; Ernst, R. R. *J. Magn. Reson.* **1981**, *43*, 259.
48. Bodenhausen, G.; Wagner, G.; Rance, M.; Sørensen, O. W.; Wüthrich, K.; Ernst, R. R. *J. Magn. Res.* **1984**, *59*, 542.
49. Kumar, A.; Wagner, G.; Ernst, R. R.; Wüthrich, K. *J. Am. Chem. Soc.* **1981**, *103*, 3654.
50. Macura, S.; Wüthrich, K.; Ernst, R. R. *J. Magn. Reson.* **1982**, *47*, 351.
51. Hull, W. E.; Sykes, B. D. *J. Chem. Phys.* **1975**, *63*, 867.
52. Wagner, G.; Kumar, A.; Wüthrich, K. *Eur. J. Biochem.* **1981**, *114*, 375.
53. Kessler, H.; Kogler, H. *Justus Liebigs Ann. Chem.* **1983**, 316.
54. Martin, M. L.; Martin, G. J.; Delpuech, J.-J. "Practical NMR Spectroscopy"; Heyden: London, 1980.
55. Abu Khaled, M.; Watkins, C. L. *J. Am. Chem. Soc.* **1983**, *105*, 3363.
56. Loosli, H. R.; Kessler, H.; Oschkinat, H.; Weber, H. P.; Petcher, T. J.; Widmer, A. *Helv. Chim. Acta* **1985**, *68*, 682.
57. Deslauriers, R.; Smith, I. C. P. In "Topics in Carbon-13 NMR Spectroscopy", Levy, G. C., Ed.; Wiley: New York, 1976; Vol. 2.
58. Wüthrich, K. "NMR in Biological Research: Peptides and Proteins", North Holland: Amsterdam, 1976.
59. Wider, G.; Lee, K. H.; Wüthrich, K. *J. Mol. Biol.* **1982**, *155*, 367.
60. Kessler, H.; Bermel, W.; Krack, G.; Bats, J. W.; Fuess, H.; Hull, W. E. *Chem. Ber.* **1983**, *116*, 3164.
61. Rance, M.; Sørensen, O. W.; Bodenhausen, G.; Wagner, G.; Ernst, R. R.; Wüthrich, K. *Biochem. Biophys. Res. Comm.* **1983**, *117*, 479.
62. Oschkinat, H. Diplom Thesis, J. W. Goethe-Universität, Frankfurt, 1983.
63. Braunschweiler, L.; Ernst, R. R. *J. Magn. Reson.* **1983**, *53*, 521.
64. Damm, I. Dissertation, J. W. Goethe-Universität, Frankfurt, 1984.
65. Kessler, H.; Hölzemann, G.; Zechel, C. *Int. J. Pept. Protein Chem.* **1985**, *25*, 267.
66. Kessler, H.; Schuck, R.; Siegmeier, R. *J. Am. Chem. Soc.* **1982**, *104*, 4486.
67. Wokaun, A.; Ernst, R. R. *Chem. Phys. Lett.* **1977**, *52*, 407.
68. Bodenhausen, G. *Progr. Nucl. Magn. Reson. Spectrosc.* **1982**, *14*, 137.
69. Braunschweiler, L.; Bodenhausen, G.; Ernst, R. R. *Mol. Phys.* **1983**, *48*, 535.
70. Mareci, T. H.; Freeman, R. *J. Magn. Reson.* **1983**, *51*, 531.
71. Experimentally it was observed that an optimization to higher coupling constants than the theoretical values yields fewer artifacts (see ref. 41).
72. Wagner, G.; Zuiderweg, E. R. P. *Biochem. Biophys. Res. Commun.* **1983**, *113*, 854.
73. Boyd, J.; Dobson, C. M.; Redfield, G. *J. Magn. Reson.* **1983**, *55*, 170.
74. Bodenhausen, G.; Freeman, R. *J. Magn. Reson.* **1977**, *28*, 471.
75. Freeman, R.; Morris, G. A. *J. Chem. Soc., Chem. Commun.*, **1978**, 684.
76. Kessler, H.; Hehlein, W.; Schuck, R. *J. Am. Chem. Soc.* **1982**, *104*, 4534.
77. Kessler, H.; Eiermann, V. *Tetrahedron Lett.* **1982**, 4689.
78. Kessler, H.; Schuck, R. In "Peptides 1982", Bláha, K.; Malon, P., Eds.; de Gruyter: Berlin, 1983.
79. Kessler, H.; Bernd, M.; Kogler, H.; Zarbock, J.; Sørensen, O. W.; Bodenhausen, G.; Ernst, R. R. *J. Am. Chem. Soc.* **1983**, *105*, 6944.
80. Bax, A. *J. Magn. Reson.* **1983**, *53*, 517.
81. Kessler, H.; Griesinger, C.; Zarbock, J.; Loosli, H. R. *J. Magn. Reson.* **1984**, *57*, 331.
82. Kessler, H.; Griesinger, C.; Lautz, J. *Angew. Chem.* **1984**, *96*, 434. *Angew. Chem. Int. Ed. Engl.* **1984**, *23*, 444.
83. Kessler, H.; Bermel, W.; Griesinger, C. Paper presented at the 25th Experimental NMR Conference, Wilmington, Del., April 9–12, 1984.
84. Kogler, H.; Sørensen, O. W.; Bodenhausen, G.; Ernst, R. R. *J. Magn. Reson.* **1983**, *55*, 157.
85. Bolton, P. H.; Bodenhausen, G. *J. Magn. Reson.* **1982**, *46*, 306.
86. Bax, A. *J. Magn. Reson.* **1983**, *53*, 149.
87. Kessler, H. *Relax. Times*, **1984**, *4*, 8.
88. Bax, A.; Freeman, R.; Frenkiel, T. A.; Levitt, M. H. *J. Magn. Reson.* **1981**, *43*, 478.
89. Freeman, R.; Frenkiel, T. A. *J. Am. Chem. Soc.* **1982**, *104*, 5545.
90. Davies, D. B.; Abu Khaled, M.; Urry, D. W. *J. Chem. Soc., Perkin Trans.* II, **1977**, 1294.
91. Barfield, M.; Al-Obeidi, F. A.; Hruby, V. J.; Walter, S. R. *J. Am. Chem. Soc.* **1982**, *104*, 3302.
92. Kessler, H. Meeting of the "Fachgruppe Magnetische Resonanzspektroskopie" of the "Gesellschaft Deutscher Chemiker", Walberberg, October 5–8, 1983.
93. Wynants, C.; Hallenga, K.; van Binst, G.; Michel, A.; Zanen, J. *J. Magn. Reson.* **1984**, *57*, 93.

94. Jeener, J.; Meier, B. H.; Bachmann, P.; Ernst, R. R. *J. Chem. Phys.* **1979**, *71*, 4546.
95. Huang, Y.; Macura, S.; Ernst, R. R. *J. Am. Chem. Soc.* **1981**, *103*, 5327.
96. Kessler, H.; Hölzemann, G.; Geiger, R. In "Peptides—Structure and Function", Hruby, V. J.; Rich, D. H., Eds.; Pearce Chemical Company: Rockford, Ill., 1983; p. 295.
97. Campbell, I. D.; Dobson, C. M.; Ratcliffe, R. G.; Williams, R. J. P. *J. Magn. Reson.* **1978**, *29*, 397.
98. Kessler, H.; Oschkinat, H. *Angew. Chem.* **1985**, *97*, in print.
99. Kessler, H.; Müller, A.; Oschkinat, H. *Magn. Res. Chem.* **1985**, *23*, in print.
100. Wynants, C.; Van Binst, G.; Loosli, H. R. *Int. J. Peptide Protein Res.* **1985**, *25*, 608.
101. States, D. J.; Haberkorn, R. A.; Ruben, D. J. *J. Magn. Res.* **1982**, *48*, 286.

Author Index*

* Each author's name is followed by the reference number. The numbers in parentheses represent the pages on which that reference number is cited.

Subject Index